The Overlook Guide to Growing
Rare and Exotic Plants

The Overlook Guide to Growing

Rare and Exotic Plants

Raymond Foster

Drawings by Rosemary Wise

THE OVERLOOK PRESS

Woodstock • New York

First published in 1985 by
The Overlook Press
Lewis Hollow Road
Woodstock, New York 12498

Library of Congress Cataloging in Publication Data

Foster, Raymond .
    The overlook guide to growing rare and exotic plants.

    Includes index.
1. Gardening.   2. Plants, Ornamental.   3. Fruit-
culture.   4. Tropical plants.   I. Title.   II. Title;
Rare and exotic plants.   III. Title: Exotic plants.
SB453.F64   1985   635.9   83-22057
ISBN 0-89751-968-1

CONTENTS

ACKNOWLEDGEMENTS

During the writing of this book, valuable help and advice was given by Susan Archer and William Davidson in the UK, and by Ross H. Gast and Frank Simerley in the USA. In addition, reference has been made to the works of numerous authors, notably E. F. Allen, H. Baker, Iris Bannochie, Beatrice Drewe, Sima Eliovson, B. Fearn, E. G. Gilbert, G. A. C. Herklots, Lucie Kofler, A. P. Legge, Max Pettersson, J. W. Purseglove, T. C. Rochford, K. Showler, Una van der Spuy, C. M. Taylor, Graham Thomas, M. Vogts and Miriam P. de Vos.

Photographs by courtesy of The Iris Hardwick Library of Photographs; Harry Smith Horticultural Photographic Collection; Michael Warren

1
RARE, EXOTIC AND DIFFICULT PLANTS

Any selection of plants linked by the common characteristic that gardeners consider them rare, exotic or difficult, is likely to show a personal bias. Originating as these plants do in so many and such varied regions of the world, it is equally probable that few of them would be considered rare or exotic within their home territory, and few also might be considered difficult in areas that are very similar to their native habitat in matters of climate, soil and site. Given the conditions they need, most of the selection in this book will grow freely outdoors in the subtropical climatic zones, in places such as Florida and southern California, and the Australian coastal towns, also in favoured but locally variable climates such as those of South Africa and New Zealand. Some of these plants are quite hardy, but because of their unusual properties need a special technique or sympathetic treatment to enable them to thrive, as for example some of the daphnes and mistletoes. Some, like the South African bulbous plants, often have a cultivated variety or local strain hardier than the norm, or one with a flowering season that differs from the common type, and is better able to cope with temperate climates. Others, like some of the gorgeous South American climbers and several shrubby Australian subjects, can withstand far harsher conditions than is often realised. Even the more tender subjects will grow and produce of their best within temperate regions of North America or Europe, including Great Britain, though it may be only with the protection of a cool greenhouse, or confined in a container which can be carried outside for the summer. All, however, need individual attention and special conditions which even the best gardeners are apt sometimes to misconstrue.

Many plants are rare simply because they find themselves at the extreme limit of their natural range – a survival zone which is decided by a combination of factors, mainly climatic, affecting the temperature and rainfall, and edaphic, concerning the type and constitution of the soil. A few areas, such as deserts, seas and swamps, are so clearly demarcated that they allow for scarcely any intermediate zone; but where most plants are concerned the territorial overlap that exists between friendly and hostile environments is considerable. This overlap between areas in which a plant can flourish successfully, and those which preclude all hope of survival, constitutes the region of half hardiness, a wide zone in which the subject can often be established outdoors under the controlled environment of a garden – a place where the soil, the micro-climate in terms of aspect and shade, and the danger of unfair competition from other plants, can all be adjusted in the adopted plant's favour.

The process of reproduction outside a plant's normal territory is often inhibited enough to prevent any natural spread. There may be insufficient warmth from the sun, or the daylight hours may be too short to ripen the season's growth, to open the flowers or to allow the seed to set. Once established in its adopted home a plant might prove perfectly hardy and vigorous, but owing to an unaccustomed climatic pattern the flowers may open perhaps too early in the spring to allow the safe production of fruit, leaving them vulnerable to severe frost; or too early, perhaps, for the bees and flies essential to effect natural pollination – as sometimes happens to the apricot within many north temperate zones. Or, though the plant might be able to grow outdoors on an annual basis, flowering and setting fruit freely, the summer season may not be long or warm enough to enable the seed to ripen properly – as is often the case with aubergine and tomato. Conversely, the winter may be too short or too mild; plants from the far north are inherently accustomed to a long period of dormancy, and when introduced to the milder temperate regions find it impossible to flourish there. The Siberian larch, for example, makes a fine timber tree which is said to tolerate temperatures as low as − 70°C (− 94°F) during its native winters, yet if planted in balmy England the same species, tricked into breaking dormancy too soon year after year, suffers severe frosting each spring and at best can make only a poor, stunted bush. To take another example from the coniferous forests of the world, the Sitka spruce from North America is the most commonly planted timber tree in the British Isles, because it grows so well on the exposed upland moors normally available for British forestry, and millions of its seedlings are produced annually in British nurseries, using imported seed; yet it is unable to spread of its own volition, and if left to its own devices the species would very quickly die out in Britain.

While the rareness of a particular plant is often the result purely of local phenomena connected with climate and soil, a few particularly attractive subjects whose habitat is fairly limited have become rare in the wild apparently through over-collection by gardeners – as for example the lady's slipper orchids of North America, Europe and Britain. Most other ground orchids are natives not of the woodland like lady's slipper, but of open grassland and marsh, and are becoming scarce only because their habitat is daily shrinking as new land is brought under cultivation, and stretches of marshland are drained. That type of rareness which has been caused by the lack of suitable habitat always supplies its own clues for a plant's cultivation. Unless their environment is extremely specialised, plants under natural conditions normally live in the close company of other plants, often in fierce competition with them; but the members of an established plant community within any particular habitat will have evolved their own levels of co-existence ensuring the survival of the whole. A woodland community, being easily observed, can best illustrate this functioning on different strata: here the tallest, most light-demanding trees dominate the others to form an uppermost canopy, while smaller trees able to bear shade grow successfully as a second storey. Beneath their combined foliage the shade-bearing shrubs and herbs flourish, and beneath and around these, a rich population of mosses and other flowerless plants. In forest clearings and around the woodland edge, light-demanding shrubs and herbs crowd into the sun and jostle for flowering space, while the woodland climbers scramble upwards often through semi-darkness, twining or sprawling, until they reach the sunlight above the canopy of foliage and are able to produce flowers and fruit of their own. In shade which is particularly dense, the ground-hugging plants often have parasitic or saprophytic natures, and thus are able to manage without the aid of sunlight. To this last group belong the woodland mushrooms and toadstools.

Unlike the fungi, plants which use chlorophyll share a common need for sunlight, for they manufacture their food from water and air through the process of photosynthesis, and different plants have become adapted to utilise varying degrees of light. On the one extreme plants which normally grow in full sunlight – especially those which inhabit mountain and desert where the sun is at its greatest intensity – cannot manage well with much less than the maximum.

A few desert plants which to protect themselves from their harsh environment grow partly hidden beneath the ground, such as lithops, the so-called stone plants, actually have translucent windows in their visible parts – the leaf tips – which allow light to penetrate their internal, chlorophyll-bearing cells. At the other extreme, as we have already seen, are the ground layer plants on the forest floor. These latter adopt a variety of subterfuges designed to make the most of their environment: some, such as the bugle, the ground ivy, the wood sorrel and the white dead nettle, grow new foliage in the autumn as the deciduous trees above them begin to lose their leaves, and remain green and active over winter in order to build up their reserves of food; others, such as the winter aconite, the celandine, the bluebell and, sometimes, the primrose and sweet violet, grow and flower in the early spring before the forest leaves have fully expanded, their foliage thickening and covering the ground until the overhead shade finally becomes too dense, and they withdraw into dormancy until the main growing season is over.

The wide variety of life styles which plants have adopted enable them to fit into any appropriate community. Below ground as well as above, each species occupies its own level, and its roots habitually draw their sustenance from a type of soil perhaps quite different to that penetrated by its deeper or shallower neighbours, for soils beneath natural plant communities often have a complex structure, ranging in depth from a dark layer of almost pure organic material near the surface, through variously coloured layers of deposition to the basic wholly mineral soil beneath. The mineral soil itself varies widely not only in mechanical composition but also in chemical constitution, and in its acidity or alkalinity. Cultivation destroys this delicately balanced arrangement and creates an amalgam in the topsoil, often allowing the least specialised plant species to adopt the status of weeds.

Bulbous plants are often those which have to cope in their native habitat with a difficult season of extreme shade, dryness or heat, for they have developed the capacity to store food and water supplies as they become available, and to draw on these reserves later in the year. Plants such as these normally continue their eccentric seasonal habits when grown under cultivation in surroundings which may be quite different. Aquatic plants, which are adapted to cope with and transpire large quantities of surplus water, can often survive on dry land provided the surrounding plant competition is not too fierce. At the opposite extreme, resistance to drought has been carried to a fine art in the case of the stone plants, the cacti and the desert euphorbias; but in plants which lack the ability to cope with varying water supplies, dormancy can often prove effective as a means of survival during difficult seasons.

Adopting its most familiar guise during winter among the deciduous broadleaf plants within temperate zones, annual dormancy can also be found in tropical woodlands wherever the seasons vary between wet and dry extremes, the trees shedding their leaves for the duration of the dry season, though the heat may be intense. These same trees may well remain evergreen if transferred to a temperate climate, and this type of dormancy can sometimes be induced in tender evergreen plants by allowing them to dry out during the autumn so that they lose their leaves, and are thus able to tolerate low winter temperatures. Plants of those tropical zones which experience little difference between seasons and are thus able to grow evenly through-out the year are, as a rule, broadleaved and evergreen, their foliage carrying an unusually high water content. Resistance to extreme cold is largely conditional on a low water content, for when internal water freezes it damages the cell structure of a leaf or stem. Broadleaved evergreens which inhabit cold regions, such as rhododendrons and laurels, typically have rather dry, leathery leaves. In very cold climates the available growing season between the last and the first frost is often too short to allow many deciduous plants time to ripen their growth before winter, and the tendency in such regions is again towards evergreen foliage; but this time the conifers are dominant, with their narrow, dry or very resinous needle-like leaves.

In some areas the growing season may be unduly shortened by limitations of rainfall, and the adaptations evolved by the native flora are often similar whether the limiting factor is one of temperature or moisture. Thus flowering annuals form the highest percentage of plants native to the desert regions of the world, where the summers are usually long, and the equally protracted periods of drought or, in some regions, sub-zero temperatures, can be passed safely as a dry, tough-coated seed. Perennial plants predominate in colder, damper regions such as the Arctic, with its short growing season; hence most herbaceous perennials under cultivation are hardy, while so many of the garden annuals are tender or only half hardy, yet well able to germinate, grow, flower, shed their own seed and finally die a natural death, all in the space of one temperate summer.

Because frost damage to a plant is largely a mechanical matter of water freezing internally, the danger of frosting at the roots is never so severe when the soil is dry. Even some tropical plants are able to survive sub-zero temperatures at the roots, provided they are kept dry. Wet soils, on the other hand, can prove fatal during the winter, even though the temperature remains well above freezing point. Snow is often a significant factor in ensuring winter dryness, provided the ground was dry in the autumn at the time of the first snowfall of the year. Unfortunately, this never happens in humid, temperate climates where snow seldom persists. The altitude of a plant's natural habitat is sometimes relevant when planning over-winter treatment under cultivation, particularly in the case of Alpine plants, for the more typical within this highly specialised group need to experience the comparative dryness that Alpine snow ensures, both at the roots and on the foliage, during their winter rest. At a lower altitude many mountainous districts are notoriously wet, and autumn rain saturates the ground which remains sodden all winter, despite a covering of snow – water-loving plants such as the rushes and sphagnum moss are typical mountain plants of the lower altitudes – but on the peaks in the Alpine plant's habitat the air is cold enough to turn the autumn rain to snow, giving the gritty mountain soil ample time to drain and dry out before it can freeze. Cold air, like water, flows downhill following the line of least resistance, so slope, too, plays an important part in preventing frost. The combined result is that Alpine plants are able to remain dry and comparatively warm until the snows melt in the spring and supply the copious moisture needed by the plants for maximum rapid growth, flowering and seeding within the short growing season available to them. Thus, even in the same region, lowlands can experience wet winters and dry summers, and mountain tops the reverse, with dry winters and wet summers.

Soil drainage is one of the most important considerations when trying to accommodate not only Alpine plants but also those which originated in a hot, dry habitat. Slope plays a part, but it is largely the composition of a soil which determines its ability to drain freely: water moves unhindered in a gritty, sandy soil; peat and similarly absorbent organic material retains moisture in the soil, and an even balance of both factors provides the free draining moisture needed by the vast majority of plants. The appropriate type of soil, coupled with details of shade and aspect that most nearly equate to the natural habitat, assessed according to the local climate, provide the basic conditions under which a plant may be grown outside its natural range. An aspect facing the sun – towards the south in the northern hemisphere, and towards the north in the southern hemisphere – is consistently the warmest site, though the temperature may well be modified by the influence of rain and cooling breezes. Facing the poles will always be the darkest, though not necessarily the coldest site. An aspect which faces the heart of a continent is often the most extreme, and liable to attract the coldest winds – such as the east winds of Britain, as compared with the prevailing mild Atlantic breezes which blow from the south-west.

The degree of tolerance to exposure a plant exhibits is also largely related to its natural habitat. Woodland plants for instance usually dislike strong winds, and exposure even to mild but continual prevailing breezes can cause topiary-like shapes due to the constant nipping back of the young shoots. The large leaves of many species can become ragged and torn, with unsightly brown edges, and woodland climbers in particular are likely to be broken. The British climbing ivy, however, is one of the hardiest and most wind-resistant of plants, and can provide extremely useful evergreen shelter and a winter windbreak in light deciduous woodland, rarely doing harm to the trees it clothes provided they are reasonably substantial. Very cold winds are particularly destructive to young evergreen foliage, and not only half hardy broadleaf plants but also conifers such as arbor-vitae and the cypresses, although otherwise quite hardy, are liable to have their foliage browned and sometimes killed. Hot or warm winds are a regular feature of some plant habitats, and several shrubs and small trees such as the South African silver tree and the Australian bottle brushes enjoy the free circulation of warm breezes, and seem to grow the better because of them.

Barriers erected to break the force of the wind in exposed gardens can be either of two very distinct kinds: permeable, or impermeable. One of the latter, for instance a wall, a solid fence, or a row of compact evergreens, can be the cause of considerable turbulence, and sometimes give rise to more damage than it prevents. A site a little to the windward can prove more sheltered than one a similar distance to the lee, so that a plant sited close against a wall but facing the wind may, in fact, be very well protected. A narrow windbreak of the former kind, to be really effective, needs to be about 50 per cent permeable, so that the force of the wind is reduced rather than diverted – the sort of sheltering cover often provided during the winter by deciduous trees, or by a mixture of evergreen and deciduous shrubs.

Exotic plants cannot always be expected to behave typically when faced with the unfamiliar characteristics of foreign plants. The local competition for light, water and soil may prove too severe – a fate that occurs in the great majority of cases when an introduced plant fails to grow; in a very few cases the new competition can prove an ineffective control, allowing the exotic plant to become a menacing weed. Although risks are inherent, therefore, in the introduction of exotic plants to new areas comparable in climate and soil, the chances of such a newcomer becoming dangerously invasive are lessened by many factors. As often as not, the exotic plants gardeners wish to grow in their gardens are on the borderline of hardiness within their particular climatic zone; most of the damage likely to be caused by new introductions has probably already been done, and considering the many thousands of exotic plants which have been distributed all over the world, dramatic results occur but rarely. A plant which bears useful fruit, like the guava, could perhaps be excused for straying from the Americas to invade countries of the Third World. Amongst the intentional introductions which went wrong were the prickly pears, which could also be described as fruit trees, though Australian farmers certainly did not thank the New World for these cacti when they saw them spreading across their ranches at the rate of a million acres a year. The beautiful water hyacinth from tropical America, originally no doubt an escapee from garden lily ponds, has found its way across the world's warm zones and choked important waterways, first in Florida, then in Australia, Africa, Indonesia, and doubtless many other countries in the five continents. A hedychium is bidding to take over the Azores, and more recently the Japanese honeysuckle, though fragrant and beautiful, has been causing problems in the USA. The cherry pie lantana was an introduction from Jamaica to South African gardens, and has been causing trouble there with farmers and foresters ever since.

The nature of the local competition which awaits an exotic plant is a powerful factor in determining the success or failure of the introduction, and the more amenable a particular habitat to plant life, the stronger as a rule will be the opposition. Only in the harsh environment of a desert can the local flora be described as free from the direct competition of other plants. In many cases the competition an introduced plant has to face can come from a lower class of plants, perhaps from a fungus, like the collar rot which attacks some succulents and the seedlings of desert shrubs and climbers such as clianthus, when grown in surroundings that would seem, on the face of it, much more favourable to healthy growth. Similarly, introduced plants may lack innate resistance to a locally prevalent disease – European elms were largely immune to Dutch elm disease before a new and virulent strain of this fungus appeared. In this connection it sometimes happens that plants whose present-day range is unusually limited, show themselves able to thrive and spread by natural means when introduced abroad – species such as the Monterey pine from California, which was planted on many thousands of acres in the moist highlands of southern Africa and acclaimed as a highly successful timber producer, before suddenly being hit and devastated by a microscopic virus disease, to which the other exotic pines in those regions were apparently immune.

The Monterey pine has a very limited natural distribution today, in common with some other North American trees, such as Monterey cypress, sequoia and sequoiadendron – all of which have now been introduced to many other parts of the world – but it seems likely that they were once of widespread distribution. Indeed, there is fossil evidence of this in the case of many species: the maidenhair tree, which as a wild plant is nowadays limited to a remote area of northern China, but during the past century has been reintroduced to many countries and is often seen as a fine street tree in Britain and North America, apparently once flourished naturally over much of the globe; and the magnolias typical now of America and parts of Asia were once, before the Ice Age, common wild trees in Europe. The most abundant plant in Europe and Britain at that time is said to have been the mountain avens, now limited in natural distribution to mountain tops and the far north – though common enough on garden rockeries.

Some of the most efficient colonisers of the plant world rank among the most ancient of life forms – the ferns. Several species of fern have a vast natural range throughout the northern hemisphere, and one, the common bracken, which has the great advantage of spreading both by spores and by tenacious underground rhizomes, is of almost world-wide occurrence from the tropics to the Arctic. Like other plants which sometimes cause problems to farmers and gardeners, bracken is one of that number which, though hardy and ancient, have not yet filled their potential range, or found their natural limit. Other invasive plants must be placed in the same category although, like the hedychiums, they may be tender and comparatively recent of origin.

Fortunately, perhaps, few plants are able to survive for long when forced to live outside their normal station, for the adaptation necessary to meet new conditions can occur only over many generations. A wishful thinker might hope to brighten the countryside by scattering a few packets of flower seeds by the wayside (a practice on which ecologists would frown), but the chances of any plants surviving to flower and seed in the face of local competition would be slim. A few plants, of course, are well equipped to meet challenges such as this, especially those 'pioneer species' which are able to spread rapidly and exert a tenacious, if temporary, hold on such potential sites as appear by chance. The fireweed or rosebay willow herb very quickly colonises woodland clearings and makes many a bright patch of pink on the scenes of recent forest fires, whence it is distributed firstly by its numerous parachuted seeds, and secondly by its fiercely competitive but shallow root system. The British foxglove makes similarly bright patches of colour in the woods and also colonises new clearings, but more gradually, by yards rather than miles, through the scattering of its countless tiny seeds catapulted when the dried spike is rattled by passing creatures. Like the many different species of thistle whose down sometimes fills the air, dandelion seeds can be wafted many miles before chancing on new sites to settle, but though pioneers they are not colonists, contenting themselves with a solitary existence, their long, heavy taproots penetrating far below the choking mat of grass roots. The birches, by their light, winged seeds, also travel rapidly and germinate quickly to put themselves in the lead as pioneer trees of new forest clearings; but, like most of the strongly light-demanding, quick-growing small trees, their stay is often temporary, while taller trees grow slowly up through the protective canopy of light foliage and, spreading their branches over the birches, ultimately kill them off and take over the site.

Not all plants are able to travel by such means, of course, and some of those with heavy seeds can extend their range only very slowly. Some streamside plants like the marsh marigold (the English kingcup and the American cowslip) eject specially buoyant seeds able to float on the water and start a new colony further downstream. The well-known phenomenon of seed distribution by berry-feeding birds is normally a fairly local process, never more obvious than when a new hedge of hawthorn, mountain ash or other berried trees and shrubs springs up unplanted along the line of a wire fence used as a perch by the local seed-eaters. It is obvious also in the areas around some tropical settlements where the guava seems to be taking over the bush (man can clearly play a part in this, too); but birds do not as a rule carry seeds across the sea from one country to another. The invasive hedychiums are helped in their spread throughout the Azores by the local rock pigeons, it is true, but the coastlines of these islands, though scattered, represent a single habitat to these cliff-dwelling birds. On migration, berry-eating birds normally fly with empty stomachs, so ornithologists tell us, and the Scandinavian fieldfare and redwing thrushes which arrive in Britain across the North Sea each autumn and fall ravenously on the British berries do not seem to bring Norwegian hawthorns and rowans with them.

Pollination depends very often on the wind, sometimes on insects, and occasionally on birds, and this too can be a factor limiting the plant's distribution, especially when local creatures are necessary for success. Several night-blooming plants rely on the co-operation of specific moths; other species from South America are dependent on humming birds, and certain southern African aloes and the crane flower are most readily pollinated by the native sunbirds reaching inside the elongated tubes with their slender bills. Though plants such as these may sometimes be carried by chance beyond their natural range, they will be unable to produce seed if the essential factor is missing from their new station.

Because of pollination difficulties, shortage of seed is often almost symptomatic of rare, exotic and difficult plants, and for this reason vegetative propagation in the garden is frequently a necessity. Cuttings of difficult plants can often be encouraged to root by the use of a spray misting system, or intermittent mist unit, coupled with soil-warming equipment which is usually thermostatically set at a temperature of 18–21°C (65–70°F). Though rooting may be achieved quite readily, difficulty is often experienced when attempting to harden off the newly rooted cuttings once they are removed from the mist. Such units should be kept open and well aired within the greenhouse, without any additional cover in the form of polythene tent or glass frame which might cause the trapped atmosphere to become harmfully damp.

Equally good and sometimes better results, with fewer subsequent losses during hardening off, can be obtained without mist by the use of soil-warming equipment of a type in which the temperature setting can be adjusted between say 15°C (60°F) and 32°C (90°F). The frame under which this apparatus is fitted is usually inside a greenhouse and loosely covered when in use with a sheet of very thin polythene film, allowed to rest on the cuttings themselves. As soon as rooting has been achieved and new top growth appears, the cuttings must be hardened off by raising the polythene in two or three stages over a week or so, before removing it completely. By using a propagating frame with higher sides, a sheet of glass can be placed horizontally over the top to serve the same purpose, and this is probably a better method in the case of very soft cuttings. The glass also should be half raised for a few days after new growth has appeared, before being removed altogether. A very effective development of these methods involves the use of perforated or scored polythene – a thin film of polythene covered with small knife cuts, which gape open only when the material is stretched. Secured over the frame, this is pulled apart at the appropriate time to admit air, and gives excellent results with soft shoot cuttings rooted over mild bottom heat.

Besides the special methods recommended for each plant, shrubby subjects in particular can often be propagated very conveniently by air layering, especially when only a few plants are required. This operation is usually carried out in the spring in the case of outdoor plants, or equally well in the autumn when the stock plants are grown under glass. The commonest method is to cut a tongue with an upward slant, directly behind a bud or at a node on the stem, and a twist of moist sphagnum moss which has been dusted with hormone rooting powder is inserted into the cut to hold it open. It is an advantage if the hormone rooting powder also contains a fungicide. A quantity of moist sphagnum moss is then arranged around the cut part of the stem to form a cylinder, and bound in place with a strip of polythene film, sealed and made airtight with adhesive tape. As new roots grow and fill the moss they can be seen through the polythene, and the stem can then be cut below the root ball, the polythene removed, and the new plant potted up.

2
BULBS AND HERBS

Aloe
Aloe arborescens; A. ciliaris; A. ferox; A. polyphylla;
A. variegata and others (Liliaceae)

Dry, sunny, rocky places in southern Africa are often brightened with the vivid red hot poker-like flowers of the aloes – in the semi-desert of Namibia, or in Lesotho high in the cold Drakensberg Mountains, or further north among the rocky pinnacles of the Chimanimani Mountains in Zimbabwe – a splash of colour on a site often inaccessible to any but the acrobatic mountain baboons, and the iridescent sunbirds which probe the tubular aloe flowers for nectar with their long curved bills.

Some of the aloes are notoriously tricky in cultivation. Often growing by nature in areas of summer rainfall where the temperature is not, as a rule, particularly high, most of them need the maximum available light, especially during the cool, cloudless days of winter. Among the easiest to grow in South Africa itself is *Aloe arborescens*, which may be increased rapidly with self-rooting basal offshoots, and spreads naturally by this means, flowering unfailingly in midwinter at the Cape, where it is liable to experience eight months or more without rain. Many of the species are of local occurrence and well adapted to their own peculiar habitat: the tall aloe *A. dichotoma* grows like a tree in the dry scrub of Namaqualand and Namibia; in the high rocky gorges at the edge of the Little Karoo, *A. comptoni* glows scarlet amongst the euphorbia bushes and mesembryanthemums; *A. pearsoni* with its brown concertina-like stems clings to the stony koppies of the Upper Karoo, and beyond these among broad valleys and arid mountains some of the hillsides are covered with the bright yellow *A. liniata muirii*; while in the high rainfall area of Inyanga in Zimbabwe, where the mountains rise to 2,500m (8,000ft), the very local *A. inyangense* and *A. saponaria* sometimes make an eye-catching splash of flame red above the grey rock outcrops.

Aloe saponaria

In the cold Drakensberg Mountains *A. polyphylla* is of very local occurrence, and protected by law. Its spike-tipped leaves form a symmetrical spiral in five tightly packed tiers, the whole making a globe up to 80cm (2½ft) across, with spikes of orange-red flowers in the spring – October in the Drakensberg. The species grows between the 2,000–2,500m (7,000–8,000ft) contours on sunny slopes facing north-west, favouring the loose, stony ground found on the scarp slope screes. The winters there in mountainous Lesotho are dry and sunny, with snow that lies for a long time on the shady slopes, but melts quickly on the sunny ones. The leaves of this most cold-resistant of the southern African aloes, heated during the day, are cooled by night to an air temperature of −6°C (21°F) or less, though the roots are probably never subjected to lower than −1°C (30°F). In the summer there are frequent thunderstorms, the water draining away quickly; but although the scarp slopes are sunny, *A. polyphylla* inhabits those crevices among the rocks which retain moisture well. If conditions become too dry for *A. polyphylla* to thrive, the leaves curl and blacken at the tips; if subjected to shade, the leaves spread out allowing the rosettes to become loose, but the plant survives and still produces flowers. Although the roots will die if they are grubbed up and exposed to the sun, the succulent leaves and stems survive and produce new roots when the rains arrive. Too much water at the roots, however, can prove fatal, inducing rot which starts in the basal leaves and spreads progressively until it reaches the growing tip. From time to time a lateral bud develops, producing a secondary rosette whose stem elongates and sprawls along the ground, enabling the rosette to take root independently if it finds a suitable spot. Lateral buds sometimes appear following damage to the growing tip, and these can be removed and rooted quite easily. For seed production cross-pollination seems to be all-important, and this is normally done by sunbirds, especially the green-plumaged malachite sunbird which is to be seen in hilly country along the eastern side of southern Africa as far north as Malawi.

Some aloes such as the spiny *A. ferox*, with heavy erect stems bearing multiple spikes of bright orange-red flowers, regularly reproduce themselves by seed, and in such species natural vegetative reproduction seldom occurs. This magnificent aloe is sometimes grown as a house-plant in temperate countries, where it can be set outside for the summer after all danger of frost has passed. It manages to survive the winter without water provided it is kept cool, but in an over-dry atmosphere often brought about by central heating it will need watering every six weeks or so, and should be grown in a very well-drained, sandy compost.

On the French Riviera *A. ciliaris* is a favourite wall plant, with spikes of tubular orange-scarlet flowers, a floppy, scrambling plant with dark green spiralling leaves. There also the easy *A. arborescens* is sometimes set to clothe high rocks, which it does with a tangled, spiny mass many paces across − one of the earliest aloes to flower outdoors in southern Europe, producing its scarlet blooms in

Aloe variegata, the partridge-breasted aloe, is one of the easiest aloes to grow from seed, and makes a fine houseplant

midwinter. In the Scilly Islands under full exposure to the Atlantic gales, *A. mitriformis* thrives but adopts a horizontally growing habit, clinging to crevices of rock and wall. It has survived air temperatures of −4°C (25°F) and flowers regularly in the summer.

Other regular flowerers among aloes when growing in temperate regions include *A. variegata*, the partridge-breasted aloe, so-called from the crescent-shaped variegations on its leaves, but a true species readily grown from seed, quite popular as an ornamental houseplant, making a squat, colourful rosette bearing foot-high spikes of pale scarlet flowers, densely surrounded by offsets; and *A. aristata*, which forms a cluster of rosettes, rapidly growing and suckering freely, bearing loose foot-high spikes of long-stemmed reddish-yellow flowers. These two do well indoors where the air is dry, and the partridge-breasted aloe in particular is averse to a damp atmosphere. Most aloes when grown in a cool greenhouse should be given as much light as possible throughout the year, and benefit from standing outside during the summer months, but both *A. variegata* and *A. aristata* enjoy a modicum of shade, and their leaves take on a purple tinge in full sunlight.

Pot-grown aloes should be repotted every two or three years, the old dead roots removed, and any offsets which have rooted independently can be potted separately at the same time. A good growing medium consists of a loam base with the addition of leaf mould, vermiculite, coarse sand and brick dust, with a little bonemeal incorporated. Drainage must be very thorough, and clay containers should be well crocked. An occasional soaking is beneficial during the summer, allowing the compost to become almost dry before watering again; but during the winter water should be withheld, provided the growing medium does not become over-dry and dusty.

In their natural state aloes hybridise very easily and may not come true from seed, and where offsets are available these form the readiest means of increase. In South Africa, aloe seeds are sown in boxes, using alternate layers of soil and peat. Germination normally occurs after six weeks or so, but except in a few species such as *A. variegata*, it is seldom successful, and vegetative propagation is usually resorted to.

Amaryllis
Belladonna lily
Amaryllis belladonna   (Amaryllidaceae)

The beautiful, sweet-scented pink trumpets of the belladonna lily appear before the leaves emerge from the bulb, each flower some 18cm (7in) across, clustered two or three, or sometimes as many as seven together, on the tip of the 60cm (2ft) long, inch-thick stem. Though closely related to the hippeastrums and nerines which are rather similar – and often incorrectly referred to as amaryllis – this South African bulbous plant is the only member of its genus.

The true amaryllis is a very popular flower in Australia where it

23

grows well, and where new strains have often been raised from seed. Several distinct varieties have been named, among them the white Hathor and Baptisii Alba, but the flowers of white forms such as these tend eventually to revert to the common pink type. An intergeneric hybrid with *Brunsvigia* has also been reared in Australia and is known as *Amaryllis* Parkeri, flowering very profusely in deep red-pink, each stem carrying as many as fifteen large flowers.

On belladonna lilies grown in the open within temperate zones the flowers appear towards the end of summer and in the early autumn; the variety Elata, a very fragrant soft pink, is particularly early. In southern England, especially if the weather is mild, the flowers are often still out in November; as they die down the leaves start to appear, and remain green throughout winter and early spring. At this vegetative stage belladonna lilies are particularly vulnerable to severe weather, and when frost threatens may be littered with dry bracken. A position amongst low shrubs can give unobtrusive shelter, and the winter-flowering *Iris unguicularis*, with its matted grass-like cover, is a useful plant to associate with amaryllis to help provide winter protection. If the foliage should be killed, the following year's flowers will be poor; but at the other extreme, it is important to ensure that the leaves are neither smothered beneath mulching material, nor overcrowded by other plants in the early spring.

In areas where the hardiness of amaryllis is in doubt, the combination of a nearby sun-reflecting wall with a sloping, well-drained site will give good results. The bulbs should be planted 15–20cm (6–8in) deep in fibrous, slightly acid, sandy loam which has been enriched with a little leaf mould and a sprinkling of bonemeal, and given plenty of water from the beginning of autumn until the leaves die down, but kept as dry as possible during the spring and early summer. Fresh bonfire ash seems to assist flowering if applied before the flower spikes appear, and a sprinkling of potassium sulphate is also beneficial, but is most readily imbibed if watered into the soil before the leaves start to wither.

Heavy rainfall can be a bar to good flowering, and in wet seasons though the district be mild the bulbs will sometimes fail completely to flower. The chief requisite for growing healthy belladonna lilies outdoors in the northern hemisphere seems to be a fairly dry summer with a reasonably high temperature, averaging at least 20°C (68°F). This equates to the dry winters typical of its native habitat, and the period during which the bulbs are able to recuperate and expand, ready

The belladonna lily, *Amaryllis belladonna*, is well worth growing outdoors where the winter temperature does not fall below −12°C (10°F)

to flower at the first rains of spring. In the South African veld where amaryllis grows wild, seasonal bush fires often clear the dry grass and other low vegetation at the close of winter, an occurrence which always seems to result in extra strong flowering and brighter flower colour − in common with other spring-flowering bulbs of the area. Fresh growth of course is more easily seen and the flowers stand out more vividly against a bare, charred background, but these fires also result in a greater proportion of the rainwater, bearing potassium-rich material, being diverted to the large bulbs just below the surface. Pasture grass throughout much of Africa is strengthened annually by controlled burning at the close of the dry season, removing the competition of woody vegetation, and bulbs, like the grass, may well benefit from fires which are not fierce enough to penetrate below ground.

In colder climates two especially robust forms of amaryllis, Rubra and Kewensis, are both popular and often grown as conservatory pot plants, and the former in particular, though somewhat variable in performance, is well worth growing outdoors as a border plant in districts where the minimum winter temperature does not as a rule fall below − 12°C (10°F), given the protection of a nearby wall and a well-drained site.

Pot-grown amaryllis bulbs normally flower well, but for a sustained performance they need careful handling. For commercial flower production the pots are usually plunged to their rims in coke breeze or similar material, laid on flat asbestos sheeting and given bottom heat of about 27°C (80°F). The glasshouse is ventilated from below, allowing the air to circulate as freely as possible and remain cool, though the ambient temperature is never allowed to drop below 4–7°C (40–45°F). During their vegetative growing period, besides regular watering, the plants are given a spray of fresh water every evening until the foliage shows signs of dying down, when they are gradually dried out and kept absolutely dry for two or three months. The bulbs are normally repotted every second year in a compost consisting of 3 parts sandy loam to 1 part well-rotted manure, allowing 1.5g/litre (2oz/bushel) of coarse bonemeal. The pots should be on the small side rather than too large for the bulbs, which are planted firmly but very shallowly − covering no more than half their depth; that is, with the top half visible above the compost. Watering should be done thoughtfully: belladonna lilies are greedy plants, and if given liquid feed when the compost is dry it will kill the foliage; plain water is best in the morning, and liquid feed when required should be limited to the afternoons.

For normal propagation of amaryllis cultivars, the old bulbs should be lifted about every five years and the small offset bulbs removed and replanted. Seed sowing will result in new, unnamed varieties, but most of the progeny will produce the common wild pink type of flower. The large seeds − about 15mm (⅝in) across − lose viability very quickly and should be sown, very shallowly, as soon as they ripen, using an evenly balanced sand/peat compost to which super-phosphate of lime has been added. Containers should be set over bottom heat of around 24°C (75°F), and the resultant seedlings potted up separately as soon as their third leaf has reached a height of some 15cm (6in). A suitable size of pot for amaryllis seedlings is the 10cm (4in) diameter, and the compost should contain a good percentage of well-rotted manure and some bonemeal. Even in warm climates, seedlings grown for the selection of new colour breaks are best handled under glass, and will be ready by late winter for potting on into the

12–15cm (5–6in) pots in which they are to flower. They should be allowed to grow without a break, given warmth and steady watering throughout winter, spring and summer until the following autumn, when they should be dried out and rested for their second winter. The average time from seed sowing to the appearance of the first flowers varies from twenty-four to thirty months.

Arum Lily
Trumpet lily; pig lily; vlei lily; lily of the Nile
Zantedeschia aethiopica (Araceae)

The African arum lilies are marsh-loving perennial herbs with the well-known cornucopian flowers which range in colour from white through shades of yellow and pink to orange and crimson. In the Cape Province, where *Zantedeschia aethiopica* is a common wild flower, the land on which it grows may frequently be flooded or frosted in the winter, and baked hard and dry for the summer. Elsewhere in the high-veld it grows on hill slopes often sheltered by bracken and brambles, and in locally temperate areas such as these may experience not only frost but even snow from time to time during the South African winter, which there lasts from May to October. Southern Africa embraces several distinct climatic zones, and the hardiest garden forms of the arum lily are likely to be those which originated in areas with seasonal features similar to the climate of their adopted home. In parts of the sub-continent which experience summer rainfall and winter drought, arum lilies start flowering in the spring and remain in bloom longer than they do in the Cape itself.

Many of the South African streamside marshes where arum lilies grow wild are subject to superficial icing, and when grown as marginal aquatic plants in temperate countries they are usually adequately hardy, provided the tubers have been set deeply enough beneath the surface to escape the winter's frost. In ornamental lakes and garden pools where the mean water level varies between summer and winter – and winter rains may raise, or summer suns lower, the surface by as much as 30cm (1ft) or sometimes more – planting should preferably be at the minimum, or dry summer water level, so that the tubers will then be submerged deeply enough during the winter to escape damage from ice. For convenience they can be established in large pots which may either be set permanently about 15cm (6in) below water level, or raised and lowered according to the vagaries of the season. Deeply set plants tend to be smaller, with fewer leaves than those planted shallowly.

In temperate regions the main flowering period during the latter half of summer is usually followed by a second flush of smaller flowers later in the autumn. At this second flowering stage the plants may be divided and the portions replanted in time for them to become re-established before winter sets in. In cold districts where heavy, deeply penetrating frosts are expected, they may be lifted in the autumn and potted up into 12–15cm (5–6in) pots, taking care not to damage the roots, and stored under glass until growth starts in the spring. In this case they may be divided at the time of planting out.

A strong compost consisting of turfy loam with the addition of well-rotted manure suits the plants very well. Manure, however, encourages algae which discolour pond water, and if they are to be planted at the water's edge, hoof and horn or bonemeal should be used instead. In drier ground, any reasonably good soil enriched with farm-yard manure will give excellent results.

As naturalised plants, in Madeira and the Canary Islands arum lilies grow wild wherever the ground is moist, sometimes forming great clumps in the valleys. In Mediterranean Europe where their distribution is governed even more strongly by the available moisture, they have become established in Italy where, in the formerly marshland district of Campagna, great drifts are to be seen growing along the stream banks. In colder northern climates arum lilies grown on dry land will survive outdoors in southern England, and hold their own even as far north as Leeds where at least one healthy clump huddles for winter warmth against a greenhouse wall. Under these conditions they need a moist but fairly well-drained gravelly soil in full sun, and winter protection ought to be given from the time of the first frost, in the form of a mulch – a thick layer of coarse peat, perhaps, held in place with a sheet of polythene – to be removed in the spring as soon as general growth starts, but ready to be replaced temporarily if a late frost should threaten.

A robust form of Z. aethiopica known as Crowborough is happy either under marsh conditions or in an open border with heavy or light soil, and is perfectly hardy in the English home counties, where it comes into flower soon after midsummer and, in mild seasons, is still flowering on waist-high stems by the time the first frosts arrive. In exceptionally severe winters it can be given a precautionary mulch of bracken, and newly acquired plants are best kept under glass for their first winter at least, young offsets being planted out in the spring after the final frost. Even after harsh winters, Crowborough sends out plentiful offsets or suckers in the spring, and these can be removed at any time of the year for propagation – preferably in the spring so as to give the new plants a whole year in which to establish themselves before flowering early the following season. If not to be planted out immediately, offsets are best potted singly in a compost of equal parts loam and peat, with the addition of a little well-decayed manure, and can be brought forward by placing the pots over bottom heat of about 15°C (60°F), leaving the tops cool and open, until they are growing away strongly.

Both in Europe and America there is a fair demand through the florists' trade for arum lilies at Easter time. They may be forced into early flower by maintaining a winter temperature under glass of between 10°C (50°F) and 13°C (55°F). Feeding starts early in the New Year, using a high-nitrogen liquid feed, and great attention needs to be paid to spraying regimes guarding against both fungal and insect pests. For this purpose they are usually pot grown in a compost of 3 parts loam and 1 part sphagnum peat plus a little superphosphate, and are flowered successively for two years before dividing in midsummer and repotting the young offsets. Alternatively, extra large flowers can be produced from open-rooted plants spaced at 25 × 30cm (10 × 12in) in a greenhouse border; or a combination of both methods may be used, growing them in pots for their first year and planting them out in the greenhouse border for the second year. Heavy flowering of such intensity quickly exhausts the old tubers, and they are normally discarded after two years of almost continuous production.

Crimson Flag
Kaffir lily
Schizostylis coccinea   (Iridaceae)

Like a slender scarlet or crimson gladiolus, in its native Africa the crimson flag usually grows at fairly high elevations, frequently along the banks of streams. Elsewhere, it should be given a humus-rich soil which does not dry out during the summer, and yet remains free draining in the winter. Being a mountain plant it is not too tender for most temperate zones, and grows well as far north as Edinburgh, Scotland, and even in the colder, drier north-east Highlands where it multiplies quickly as a garden border plant and, particularly during

Zantedeschia   aethiopica Crowborough. This robust form of the African arum lily will still be flowering on waist-high stems when the first frosts arrive

Schizostylis coccinea, the crimson flag, an African native of stream banks at high elevations, is not too tender for most temperate zones

mild winters, produces flower spikes sometimes as late as Christmas day. Further south it grows in the Botanic Gardens of the University of Liverpool across the River Mersey in the Wirral Peninsula, flowering well each autumn in company with agapanthus, and forming large clumps without winter protection in the humus-rich soil overlying deep clay – a soil which holds the moisture summer and winter; at the Royal Botanic Gardens at Kew near London it enjoys a sheltered, sunny position amongst dwarf shrubs which allow the flowering spikes to come up through their foliage in the autumn, and give them some protection during the winter; and in the Royal Horticultural Society Gardens at Wisley in Surrey the crimson flag enjoys a warm, sunny, well-drained border sheltered by a nearby wall, and grows in the company of nerines and amaryllis; it has been planted there on the rock garden, too, but in that situation early autumn frosts often spoil the flowers, though the plants themselves survive.

As a plant, therefore, crimson flag seems to be completely hardy in most parts of Britain, and often survives conditions which fall well short of the ideal; but as the flowering season starts so late in the year, the flowers themselves are very apt to suffer heavy damage from early frosts and the wet autumn weather. Even in the mild conditions of the English West Country where, in places, the crimson flag can become almost a weed, flowering is often poor; though it luxuriates in the climate, particularly when given a moist site – the bank of a stream or a peaty, semi-shaded hollow – vegetative growth tends to become over-luxuriant at the expense of flowers. Though outright drought will soon kill even the healthiest clump of plants, flowering in such places is often greatly improved if given a sunny but fairly dry site in ordinary garden soil, away from likely pockets of frost.

Plants which originate in the higher altitudinal range usually flower earlier than middle-range specimens when given comparable conditions. My own experience of crimson flag in the wild is in eastern Zimbabwe, where the typical bright-red-flowered type grows on the edges of streamside marshes around the 1,200–1,500m (4,000–5,000ft) contour, and a distinct form is found higher in the mountains between 2,000–2,250m (6,500–7,500ft) – a region in which the annual rainfall measures some 2m (6ft), falling mainly in the summer, though with mist and drizzle all the year round, and the winter temperature often falls below freezing point. In these higher regions, where the glowing crimson-pink flags contrast with the pale lilac and grey of the local *Gladiolus flexuosus* with which it keeps company

31

among the low mountain vegetation, the underlying rock strata sometimes trap water in the thin soil after heavy rain, and in these places
the plants grow and come into flower, very often with their roots
immersed in these puddles. An exceptionally fine-flowered specimen
of this mountain form, introduced to a lowland garden, consistently
flowered considerably earlier than the type, and has been given the
cultivar name Tambara. Its early flowering habit has proved very
valuable in temperate countries, especially in Britain where other
choice cultivars of the species are so often spoiled by the earliest
autumn frosts.

Tambara gives its best results if divided every second or third year
after flowering, in the autumn. If divided in the spring as is usually
recommended for the species, the resultant setback causes it
temporarily to lose its early flowering characteristics for at least the
first year following division, and in this respect it is then no better
than the other cultivars. Tambara sets seed freely and is said to
reproduce true by this method, but like all the crimson flag varieties it
can be increased quite rapidly by vegetative means, though its underground spread is not as vigorous as the sometimes invasive bright pink
Viscountess Byng, which often needs greenhouse protection for its
late flowers. Other large-flowered clones are the bluish-toned Tom
Barnard and the vigorous Major, which has blood red blooms some
6cm (2½in) across. Tambara has proved hardy in Britain, even during
the worst winter of the century. It flowers a full month earlier than
the favourite Mrs Hegarty, which itself is rather earlier than the true
S. coccinea. Last of the cultivars to bloom is Viscountess Byng, which
carries the flowering season through until the winter.

Cypripedium
Lady's slipper orchid; mocassin flower
Cypripedium calceolus; C. reginae and others   (Orchidaceae)

The European *Cypripedium calceolus* is a much prized woodland ground
orchid with drooping moustached, broad-lipped flowers on foot-high
stems, the yellow lip, with dark red blotches inside the throat, and the
maroon wings and hood contrasting warmly with the bright
yellowish-green of foliage and spathe. A rare and protected plant in
many parts of Europe; a favourite mountain woodland rarity in
Germany; and a treasured wild flower as far south and east as Greece,
where it is equally scarce, blooming high in the mountains at the sub-

alpine level. In England it still survives as a rare, protected native in one or two areas, particularly a few isolated woods in Yorkshire where the soil overlies limestone – the natural woodland home of the British ash tree – where it grows in small pockets of humus-rich soil derived from the annual leaf fall accumulating in the moist hollows; where the spent leaves themselves are slightly alkaline, but the high rainfall gradually leaches out most of the lime until the soil shows a neutral or even slightly acid pH reaction.

As a garden plant the European lady's slipper sometimes seems to thrive in an acid soil, provided the site is well shaded, and for this reason has often been thought to shun lime; but the ideal home for it is within a humus-rich hollow, surrounded perhaps by calcium-bearing material, and in peaty garden beds in limestone areas, such as the Cotswold hills in southern England where, although the water used for summer irrigation may be very hard, it often grows to perfection. Even in areas with alkaline soil where no acid site has been specially prepared, *C. calceolus* will grow well if the other conditions are suitable. In one beautiful Suffolk garden where the soil has been washed neutral near the bank of a stream, though surrounded and overlooked by alkaline land, the lady's slipper thrives beneath the shelter of an old tree stump, shaded, supported and backed by clumps of the hardy maidenhair fern. An equally promising site might be found at the foot of a rock facing away from the winter sun, finding a spot where the moisture will percolate slowly and prevent drying out during the summer months. Nevertheless, the soil on such a site must be sufficiently free draining to prevent any suggestion of water-logging. Though not by nature a waterside plant, a shaded position on the banks of a stream or larger river often proves ideal – as near the River Cam where it flows through a university garden at Cambridge, England, and where a notable clump of lady's slipper has regularly produced two dozen flowers at a time, surrounded by snake's-head fritillaries, *Fritillaria meleagris*, and sheltered by long grass – a favourite spot, incidentally, for slugs which do a great deal of damage to the orchid crowns whilst they are still dormant.

C. reginae, the North American mocassin flower or showy lady's slipper, is less well suited to a limy soil and meets with the greatest success when grown in boggy woodland conditions without a limestone base. As beautiful as any of its genus, the petals and sepals are pure white, the pouch basically white but dotted and suffused with magenta-purple, set on 45cm (18in) stems amid light-green foliage.

Fritillaria meleagris – an exquisite companion for the lady's slipper orchid

Found in the wild over a wide range of eastern Canada and some northern states, at least as far south as Philadelphia, the mocassin flower is nowhere plentiful and, like *C. calceolus* in temperate Eurasia, has been collected so extensively in the past that its status in many places has become endangered. In common with the other species, mocassin flowers taken from the wild often fail to thrive when transplanted to a garden, seldom surviving more than a year or two. Almost invariably, the cause of this failure should be sought in the habitat: these woodland orchids need a rich and porous soil in which the roots can become firmly enmeshed among decaying twigs. The lack of success in introducing these plants to the garden has sometimes been blamed on an unidentified mycorrhizal association that might be lacking in garden soil – a situation familiar to growers of heaths and heathers – but any fungal organisms of this nature would be introduced with the roots of an imported plant and automatically injected into the new soil, and the blame is unlikely to lie in this direction. The special site requirements of the mocassin flower are plain to see in Ontario where these orchids grow in the damp lakeside woods; there the snow lies annually until April before it thaws, and the plants start to grow away in May, their roots often still submerged in the flood water that has filled the small hollows which they inhabit.

An ideal home for *C. reginae* and the other North American cypripediums, such as the fragrant maroon and yellow *C. parviflorum*, and the sulphur and pink *C. acaule*, can often be constructed by removing small pockets of the existing garden soil and substituting a compost consisting of 2 parts coarsely fibrous sedge peat and 1 part sphagnum peat to 1 part coarse grit, with the addition of bonemeal, surfaced and lightly injected with a deep mulch of well-rotted twiggy leaf mould. The site for these woodland ground orchids must be shaded by shrubs, or at least on the sunless side of a retaining wall or bank so that they receive direct sunshine only during the evenings of summer months. The Himalayan lady's slipper *C. cordigerum*, with a pure white tongue and greenish yellow petals and sepals, also appreciates sunless conditions for, lacking the protective late snow cover of its native mountains, if planted in too warm a spot it will sometimes start growth too early in the spring and suffer damage from late frosts.

Many of the species are difficult to raise from seed, and the seedlings grow only very slowly. *C. calceolus* for instance is said to take thirteen years to come into flower after germination. Once well established, lady's slippers can be divided just as growth starts in the spring, and this offers the most convenient method of propagation.

Hedychium
Ginger plant; wild ginger
Hedychium densiflorum; H. gardnerianum and others   (Zingiberaceae)

The hedychiums are somewhat similar to the cannas in their tropical appearance, with large, upright, pointed leaves and eyecatching spikes of often sweetly scented flowers in red, orange, yellow and white. Like the cannas, they are eminently suitable subjects for tropical bedding schemes within temperate zones, or as exotic tub-plants for seasonal patio display, but most are hardier than the cannas and able to survive a modicum of frost. During their growing season they need a moderately rich soil, and ample water supplies enriched occasionally with liquid fertiliser to keep them lush and healthy. A few species are winter hardy in southern England, and given some protection will survive outdoors in Britain as far north and east as Cambridge; but under these borderline conditions they lose their normally evergreen leaves and adopt a deciduous habit. To enable them to flower really well in regions such as this, they need a warm spring followed by a hot summer, with sufficient water and extra food given them to replace the nutriments lost with their leaves.

The lateness of their flowering means that early autumn frosts sometimes intervene outdoors to prevent the full display developing. *Hedychium greenei* from Bhutan produces its 5cm (2in) tubular scarlet flowers during autumn and winter under heated glass, though there is a form of the species with creamy yellow flowers which normally open in the late summer. The pale yellow *H. gardnerianum*, the sweet-scented white *H. coronarium*, and the brilliant red *H. coccineum*, all from India, are possible subjects for outdoor use in temperate climates if given some winter protection, but they do better when pot grown within an unheated greenhouse. Also from India, *H. spicatum* is hardier than these, but its yellow flowers are not so attractive. *H. densiflorum*, with small dense spikes of orange flowers, originates in the mountains of Nepal and Assam, and being naturally deciduous, though the leaves are quickly cut back by frost, is hardier than the evergreen species. It has been grown for many years in southern English gardens, where it will withstand air temperatures as low as −9°C (15°F) without demur, and makes a vigorous clump-former during spring and summer, enjoying a well-drained, loamy soil. There seem to be two forms of the species, corresponding to the upland and lowland specimens found in the wild, the strain which originates in the higher mountain ranges being the hardier. In the English counties to the north of London this hardier form of *H. densiflorum* grows well in heavy clay if given a south-facing site on a hill slope, or a raised bed to prevent waterlogging. Some of the best outdoor specimens in the London area receive little winter protection besides their own dead stalks, but this litter may well be augmented with bracken or a few

36

Hedychium gardnerianum. This pale yellow flowered wild ginger plant from India is able to grow in mild temperate areas, and sometimes becomes invasive

evergreen branches strewn on top if heavy frost threatens; in this way, firmly established specimens were able to survive the notoriously hard winter of 1962–63. Being late to emerge from the ground, the first leaves seldom showing before early May, they usually manage to escape the destructive late spring frosts, a good clump producing two or three dozen flower spikes annually, the pastel shade of pinkish apricot-orange showing up well against the fresh dark green of the 30cm (12in) leaves.

H. gardnerianum is quite content with the climate in Cornwall where, in one or two places, large free-flowering colonies exist, the mild autumns of south-west England allowing the time needed for the flowering spikes to beat the onset of wintry weather. The species was introduced to the Azores midway through the nineteenth century, and in those islands the mild Atlantic breezes, the humidity, the mean temperature and the soil have combined to provide over-ideal conditions, allowing it to become a widespread weed and a great nuisance. In the Azores frost at sea level is unknown, and although the air remains constantly moist, rainfall is relatively low – the precipitation at São Miguel averaging only half that of London, while the mean minimum temperature is some 5°C (10°F) higher. This wild ginger, in 1870 confined modestly to São Miguel gardens, was noted as establishing itself here and there by the turn of the century as a wild, but still fairly rare and local plant. Today, its tough rhizomatous roots form a solid web over miles of good land, choking crops, blocking streams, and clinging tenaciously even to rocky hillsides, where the thin soil in places is covered with a thicket of banana-like leaves some 1.5m (5ft) tall. The spiky yellow and apricot flowers are less showy than most of the hedychiums, but fragrant enough in October to lend a spice to the Azorean air. The coral-red seeds are continually being distributed by birds which eat the fleshy seed pods, and the established plants root themselves avidly in any sort of land, good or bad. So fierce is their growth that they even choke out young plantations of pine and wattle trees, and the strong-flying rock pigeons which breed around the coastline carry the seed to the farthest islands of the archipelago, some of which are isolated by as much as 240 kilometres (150 miles). Although they are beautiful plants, it is perhaps as well not to be the first to introduce hedychiums into new areas which have a mild, moist climate.

Wild ginger seed, if available, germinates readily in most cases, given a standard peaty seed compost and a bottom heat of about 18°C (65°F); but seed is seldom produced as readily by cool greenhouse or outdoor plants within temperate zones, as it is by *H. gardnerianum* in the Azores. Division of the rhizomes is a simple and very effective way to propagate any of the species under glass, and the operation is usually best carried out in the early spring – in the case of outdoor plants, as soon as new growth starts and the danger from frost has passed.

Stone Plant
Living stones; pebble plant; lithops
Lithops spp.; *Conophytum* spp.; *Dinteranthus* spp.   (Aizoaceae)

Native to some of the driest regions of southern Africa where the rainfall can average as little as 6–12cm (2½–5in) per annum, succulent plants such as lithops have, like the cacti, evolved a shape that enables them to retain such water as becomes available for as long a period as possible, and during their growing season they may contain at least 96 per cent water. Succulence such as this makes these plants very attractive to animals, and in order to avoid being eaten they are often fiercely armed, as in the American cacti, or occasionally mimic the surrounding pebbles in appearance, as is the case with lithops and the other living stones. Most of these stone plants are very difficult to see in their native habitat, even by skilled botanists who may be searching for them. Their colour, like their size, is inherently variable so as to match the local pebbles – which in the semi-desert regions often consist of small fragments of quartzite varying by locality from white and grey to yellow and greenish, pink and brick red. Stone plants of the wrong colour or shape are very soon found and eaten by thirsty baboons or birds. During the comparatively brief rainy period whilst these plants are made more conspicuous by their flowering, the rains have normally ensured other more substantial food supplies in the area, and such small plants as lithops are then less likely to be eaten.

Lithops spp.

Despite their highly specialised existence in nature, the living stones are comparatively easy to grow in a cool greenhouse if given sympathetic attention, and can make unusual and very interesting houseplants.

The true lithops have a very distinctive shape, evenly cleft into two fleshy, rounded leaves, and beautifully mottled – when studied at close quarters – in pastel shades of grey, brown, violet or green. Under cultivation, all watering should stop after the flowers have died down in the early autumn, and any seeds which have set should be harvested then. Watering of lithops should be restarted in mid-spring, unless the winter has been particularly mild, in which case watering must be left a further month until late spring, otherwise the flowers may fail to develop. Once watering starts, the old leaves wither as new growths appear in the central cleft and continue to grow until flowering time in the late summer. The dead remains of the old leaves should be left intact to protect the new growth, especially if the plant is subjected to hot sunshine. The appearance of short-stemmed white or yellow daisy-like flowers, which also issue from the central fissure, and are often larger than the visible plant itself, marks the end of the year's vegetative growth.

Other closely related stone plants include dinteranthus, which have

These stone plants, *Dinteranthus* spp, need a clay pot deep enough to accommodate their long fleshy taproots

grey, pebble-like bodies, two-storied like a miniature cottage loaf; and conophytum, of various dimensions, often kidney or heart shaped and usually tinted an overall greyish green. *Conophytum albescens* is covered with very fine white hairs, and when grown in the greenhouse produces yellow daisy-like flowers freely over a long season between spring and autumn; *C. bilobum* is somewhat similar in appearance, readily clump-forming, with a good display of yellow flowers. When these two species finish flowering in the autumn new vegetative growth commences immediately and water should continue to be given until midwinter, at which point the plants should be rested with no further watering until mid spring. The tiny *C. minutum* has rounded grey-mottled stones and spidery pink flowers; *C. velutinum* has a heart-shaped stone body covered with short, velvety hairs, and produces showy purple flowers. Both these species need watering fairly generously for a month or so in the earliest spring, but all water must be withheld for the following two months to form a late-spring resting period. New vegetative growth begins in midsummer when water must again be given; flowering commences a month later and continues until the autumn, and all watering should cease before midwinter.

Stone plant stems are hidden below ground and spring from the thick, fleshy taproot which shrivels during the drought period, often drawing the withered leaves down below the surface of the soil as it does so. Fibrous, hair-like feeding roots sprout from the surface of the taproot as soon as water appears, and wither in the dry season, at which stage it is impossible for the plants to imbibe water through the roots. Too little water is usually a safer option than too much, as over-wet conditions can quickly cause rotting off at the root collar.

The visible parts of stone plants are in fact merely the tips of their leaves, and some species have translucent windows built into these exposed tips so as to let sunlight into the chlorophyll-producing cells deeper in the leaf and stem. Desert plants such as these need maximum sun, and when grown under glass in the restricted environment of a pot sometimes tend to suffer from scorched roots. In order to avoid this, pots containing stone plants are best kept plunged in a mixture of sand and peat. Extra long clay pots allow unrestricted rooting, and while clay pans are ideal in which to grow decorative groups or small colonies of lithops and other living stones, they must also be deep enough – 6–10cm (3–4in) – to accommodate the long taproots. Irrespective of the type of container, the visible leaves should be surrounded by carefully matched pebbles which will not only enhance the display, but also help to keep the roots cool and moist.

While the winter treatment of stone plants is similar to that appropriate for the terrestrial cacti, they need a slightly higher minimum temperature, not allowed to fall below 10°C (50°F). Throughout the winter full light is essential, and a sunny, well-ventilated greenhouse serves best; but they can be kept indoors during this period on a sunny windowsill, provided the site does not become too cold at night – they should not, for instance, be excluded by curtains from the heat of the room.

Young lithops seedlings are best grown without a break and watered moderately for a whole year, by which time they should be about the size of a small pea; but they can be transplanted if required when only six months old, using a compost which might consist of 2 parts sterilised loam to 1 part sphagnum peat, 1 part coarse sand, 1 part fine grit or shingle with crushed brick, and a little lime. During their second winter they should be rested under glass by withholding water from late autumn until late spring, recommencing watering about one month after the final frost in cold temperate zones. By springtime the old pair of leaves will have taken on a shrivelled appearance, though there will be no sign as yet of the new leaves which, at this stage, are still covered by the dead skins. In the spring one good soaking of water will enable the taproot to expand and encourage feeding roots to appear, at which point the new leaves will start to swell visibly. As the young leaves become visible more water should be given, watering thereafter about once a week, depending on the surrounding humidity and allowing the compost to become fairly dry between successive waterings. Many species commence flowering in their second year.

Seed from the living stones, as is the case with most stemless mesembryanthemums, usually germinates readily and retains its viability for many years. Some species also produce offsets, and a few others will take from cuttings. Seed should be sown in the early autumn, using clay pans, and a compost consisting of 2 parts coarse sand, 1 part sphagnum peat, and 1 part gritty loam. Sunlight is necessary for germination, and the seed should be scattered on the surface without any covering. Kept moist and given bottom heat around 15°C (60°F), germination should take place within a few days and is normally at least 80 per cent successful.

Strelitzia
Crane flower; bird of paradise; bird's tongue; queen flower; Queen Charlotte's flower
Strelitzia reginae; S. augusta (Musaceae)

Strelitzia reginae is one of the most strikingly beautiful herbaceous perennials, with its strangely shaped orange and steely purple-blue flowers, each like the head of a crested crane, set singly on long, stiff stems amid large, glaucous green banana-like leaves. A native of South Africa, it is to be seen in cultivation throughout the African continent wherever ornamental gardens are watered regularly, even in the hottest desert zones, and has also become a popular border plant in the southern USA, in Australia and elsewhere. Crane flowers are a familiar sight in places such as Madeira and the Canary Islands where both *S. reginae* and the much taller, white-flowered *S. augusta* adopt the status of shrubs rather than herbs, growing typically, intermingled with oleanders and hibiscus, beneath the flame trees which line the streets of Santa Cruz. On the French Riviera the crane flower sometimes forms great clumps, flourishing in the clayish loam soil that bakes hard during the summer beneath limestone cliffs; a plant to be appreciated from close quarters, never a mass of bloom, but seldom without a dozen or so solitary flowers.

A high temperature is not necessary for successful flowering, but it should not be allowed to drop below 10°C (50°F) at night, and where this occurs during the summer *S. reginae* can only be grown under glass. Provided it is given ample ventilation it flourishes in the greenhouse, where large tub-grown plants are seldom without flowers in the late winter and early spring, with occasional odd flowers appearing throughout the year — a favourite with flower arrangers for their exotic appearance; they last well when cut, and have a useful length of stem. Under glass, the thermometer should ideally register 18°C (65°F) from early spring onwards, and no extra heat is required once this temperature has been reached naturally. Water should be given very freely from spring to autumn, maintaining a moist atmosphere during the summer by damping the floor and staging whenever the sun shines, but during the winter the atmosphere should remain as dry as possible.

Flowering is most successful on plants which are slightly tub-bound, and they should be divided and repotted only when absolutely necessary. When grown outdoors, or in the greenhouse border, several plants clumped closely together flower better than they would if isolated. Greenhouse borders intended to accommodate strelitzias need to be well dug and fertilised, and after planting should be dressed annually in the winter with bonemeal. Crane flowers, however, take up a great deal of room in a greenhouse border and, as a rule, a tub is far more convenient: it can be carried outdoors during warm spells, provided the nights are not too cool. Crane flowers will stand a little shade in the summer, but absolutely none during the winter, especially in climatic regions which have a moist winter atmosphere.

A suitable compost for tub-grown plants might consist of 3 parts turfy loam, 2 parts sphagnum peat and 1 part coarse sand, and to every bushel, or average garden wheelbarrow load of this, should be added 40g (1½oz) ground limestone, 110g (4oz) superphosphate of lime, 110g (4oz) hoof and horn meal, and 60g (2oz) sulphate of potassium.

The peculiar structure of the crane flower allows seed to set on greenhouse plants within temperate zones only if the flowers have been hand pollinated. In the wild the flowers are visited in turn by sunbirds whose long curved bills, exploring the crane's crest for nectar, transfer pollen from stamen to stigma. If seed is required under glass the process must be simulated by probing the flowers in turn with a conveniently shaped piece of wood or an artist's paint brush, twirling it a little to pick up the grains of pollen. Seed should be gathered as soon as the capsules ripen and sown immediately, in pots containing a compost of 2 parts loam, 1 part sphagnum peat and 1 part sand, plus a little superphosphate and a sprinkling of ground lime. The pots should be watered well, set over bottom heat of 21°C (70°F), and covered preferably with scored polythene which can be stretched to admit air gradually as germination takes place. After potting on into small individual pots – underpotting is better than overpotting – seedlings should continue to receive bottom heat of about 18°C (65°F), but the tops should be kept open, and care should be taken to ensure that the compost does not become dry. Seedling strelitzias take at least five years from germination to flower, but well-established plants that have been divided, or propagated merely by removing and replanting the suckers that frequently appear, are ready to flower immediately. Vegetative methods are plainly the best when only a few plants are required. Division or sucker removal for propagation is best carried out in the spring after growth is under way, giving the newly potted portions bottom heat of 18°C (65°F) until they are well established.

Watsonia
Bugle lily
Watsonia spp.   (Iridaceae)

These handsome South African members of the iris family somewhat resemble the gladioli whose native habitat they often share, and comprise some of the prettiest wild flowers of the Cape, displaying a whole range of colours from white through shades of pink, apricot, yellow, orange, scarlet, brick red, and crimson to purple. Visitors to the Cape in the spring seldom fail to notice the bright red flowers of *Watsonia spectabilis* enlivening the countryside, growing on the hills with little brown and yellow ground orchids, while the roadside ditches are often bordered with the tall orange-red *W. tabularis concolor.*

These spring-flowering watsonias sprout new leaves in the autumn and grow actively through the South African winter; in the summer, after flowering is finished and the seed has ripened, the leaves die down, allowing the corms to remain dormant throughout the long, dry summer. When grown in regions which have wet summers, those species which originate in areas of winter rainfall are often best lifted after flowering and stored under cover to simulate the months of arid dormancy they would experience in nature. Even if left in the ground after the new leaves have died down, watsonias should be lifted every four or five years and the offset bulbs thinned and replanted.

Their natural growing medium is a light, rich, well-drained soil; many, like *W. tabularis*, grow naturally in moist places, and they all need plenty of water during the growing season. Under garden conditions a sunny, well-drained site is advisable, and a sandy loam which has been enriched with peat or leaf mould usually proves suitable.

In the Royal Botanic Gardens at Kew, London, watsonias are classified for convenience into two groups – the early and the late flowerers. The first group comprises those species that flower during the spring and early summer of the north temperate zones, and these may be lifted after the foliage has died down and stored under cover; they include the dwarf crimson-flowered *W. coccinea*, the rather dwarf scarlet *W. aletroides*, the medium-sized bright pink *W. meriana*, the tall rose-red *W. pyramidata* and the scarlet *W. fulgens* with its multiple flowers, the very tall large-flowered pink *W. versfeldii*, and the giant pink *W. marginata*. The second group of watsonias, which flower during the late summer and autumn of north temperate zones, are more or less evergreen and should be left in their permanent places;

The orange bugle lily, *Watsonia beatricis*, is an autumn flowering evergreen which should be planted out permanently on a sunny, well drained site

they include such species as the short, bright orange-red *W. galpinii* – one of the best known garden watsonias – the fairly short but many-flowered *W. densiflora*, usually bright pink but with variations ranging from white to crimson and purple, the medium-sized white-flowered *W. ardernei* and the orange-red *W. beatricis*, the tall pale pink *W. longifolia*, and the giant purple-flowered *W. wordsworthiana*.

Summer- and autumn-flowering watsonias originate chiefly in the summer-rainfall areas, and the spring flowerers are mainly from those parts of the Cape which have a Mediterranean-type climate. *W.*

transvaalensis is a high-veld plant flowering in the autumn, usually in pale pink, though selected forms are available with bright pink or pure white flowers. This species becomes fully dormant during the winter and thus escapes the risk of frost damage when grown in colder climates. The spring-flowering kinds tend to put in an appearance too early in the season when grown in temperate zones, and are then easily cut back by frost.

Many excellent watsonia hybrids have been raised in Australia where they grow well and are very popular – as indeed they are in the south-western USA. The mild Atlantic climate suits their growth, too, and they thrive in the Canary Islands and Madeira, and also in Portugal and southern Spain along with other bulbous or fleshy-rooted plants such as agapanthus, freesia, babiana, ixia and crinum. In Madeira both the spring- and autumn-flowering kinds have become naturalised as an introduced wild flower, interbreed freely, and often form extensive drifts on the hill slopes.

At Kew Gardens in southern England, *W. coccinea, W. meriana* and *W. fulgens* make a splash of colour in the early summer and, in the shelter of a warm wall, help to form an African border with belladonna lilies, nerines and flame lilies. Further north and on higher ground in English gardens, watsonias can suffer from the effects of frost and biting winds, and those which are left outdoors for the winter are best given the protection of ashes piled over the dormant corms. Although late-flowering watsonias are not considered completely hardy in the Royal Horticultural Society Gardens at Wisley in southern England, they thrive outdoors without trouble in the famous Inverewe Gardens in the Western Highlands of Scotland, making a massed display in the sunniest borders there during September.

While propagation is normally a simple matter of removing and replanting offsets, the tiny seeds usually germinate freely. They should be sown in early spring, using large pots or clay pans with a compost consisting of 2 parts coarse sand to 1 part sphagnum peat, incorporating a little superphosphate of lime, and given bottom heat of 18°C (65°F). The seedlings should be thinned only if overcrowded, otherwise they should be allowed to remain undisturbed for their first year. Hardening off should commence soon after germination and the pots stood eventually in an airy greenhouse or, if the summer is fine, outdoors. During their first winter the compost should be kept moderately dry, with a minimum temperature of 7°C (45°F).

3

CLIMBERS AND SPRAWLERS

Banksian Rose
Lady Banks's rose
Rosa banksiae (Rosaceae)

Ideal conditions for the semi-evergreen banksian roses are not always readily apparent; but they can be found in the most unlikely surroundings, and local peculiarities of micro-climate and aspect are sometimes of greater significance than the latitude of a place. A bungalow I once lived in had perfect banksian roses growing round the door; the double white and the double yellow varieties, planted by a previous resident, were both covered with flowers from early spring until well into summer, with healthy annual growth – vigorous but never rampant, each plant a convenient 3m (10ft) in height by 4.5m (15ft) in spread. Strangely enough, the setting was in tropical Africa – a site facing east towards the moist, balmy breezes which rolled off the Indian Ocean. The elevation was around 2,000m (6,500ft), and in the winter, before the sun rose in the mornings, a white rind of frost would cover the more sheltered hollows and, occasionally, a paper-thin skin of ice would seal the streamside marshes; in the summer, moist and cloudy, the temperature seldom rose much above 27°C (80°F) – the whole adding up to a local climate in my garden that supported such plants as the Australian scarlet bottle brush *Callistemon speciosus*, but was too harsh for the tender flowering gum *Eucalyptus ficifolia*; where both the subtropical *Hibiscus rosa-sinensis* and the temperate *H. syriacus* could thrive side by side; and where hybrid roses flourished as bushes, but tended to scorch up when grown in standard form. That, in my view, was the ideal situation for *Rosa banksiae*, and since leaving Africa many years ago I have inevitably grouped all intended banksian rose sites on either side of that ideal.

A rare native of woodlands within limestone regions, the lady's slipper orchid, *Cypripedium calceolus*, needs humus derived from the annual leaf fall

48

In its typical form *R. banksiae* has single white violet-scented flowers each measuring some 3cm (1¼in) across – a form given the sub-specific status of *R. b. normalis*, and said to have been first introduced to the West from its mountainous home in central China in 1796, when it was brought to Megginch Castle in Strathtay, Scotland. In the wild, it is said to grow in a rampageous tangle over trees and bushes and to form bramble-like thickets in the gullies. Though the yellow forms are usually unarmed, the white flowerers have some thorns, mainly towards the base of the stems. Perhaps the commonest form to be found in gardens is the double white Alboplena, almost globular with the fullness of its petals. This and the single yellow *R. b. lutescens* are usually fragrant, but this quality varies from plant to plant and the double yellow Lutea is barely scented.

In cool temperate areas, a wall which faces both sun and prevailing breezes usually provides the ideal site for all four forms. Banksian rose blooms are borne on sub-lateral shoots, and these must not be pruned off or the season's display will be lost; if pruning is needed to keep a specimen within bounds, it should be done directly after flowering is

Lapageria rosea, the Chilean bellflower, varies widely in colour, and some of the most attractive forms are bi-coloured with a marbled effect

51

A house wall which faces both sun and prevailing breezes provides the best site for the banksian rose. The double yellow form shown here, *Rosa banksiae* Lutea, is usually thornless

finished. Sometimes, banksian roses sited on a warm wall facing the noonday sun tend to produce vigorous but barren foliage – Lutea in particular is capable of making annual growth of some 2.5–2.7m (8–9ft); newly planted specimens should be allowed to make considerable growth without pruning, and young shoots should be tied in place in anticipation of the appearance of flowering sub-laterals after three years or so of flowerless growth, for the basal wood must be perfectly mature before flowers can be produced. As plants become older and well established their growth rate slows down, and the only pruning necessary will be an annual clean-up of old, weak and unwanted branches – an operation which will encourage strong young shoots to spring from the base.

Banksian roses can make very large plants, and a height of 9–12m (30–40ft) on a sunny wall is quite usual in temperate regions. In the famous botanical gardens at Charleston, South Carolina, almost tree-like specimens with 20cm (8in) diameter stems climb through the hackberries and sycamores. On the walls of Birr Castle in County Offaly, central Ireland, the double yellow Lutea scrambles through the branches of a winter-flowering *Buddleia auriculata* to achieve an impressive height of some 18m (60ft) – on a fairly limy soil, which suits the species well. The double yellow form also thrives on strongly alkaline soils based on chalk, and shows its satisfaction by flowering extra early each year. On house walls, where the long, arching branches are so convenient to train around doors and windows, *R. banksiae* is sometimes interplanted with other climbing roses, and the combination is healthily green and attractive – but their flowering seasons rarely coincide, for the banksian flowers have almost invariably finished by the time the other roses appear.

Propagation is best achieved with cuttings about 20cm (8in) long, taken in the autumn from firm, pencil-thick trails, and set some 15cm (6in) deep in a cold frame, there to stay until the following autumn. Alternatively, sideshoots with a heel can be taken in midsummer, and set in a shaded propagating frame. Mist treatment can be helpful in rooting soft, leafy cuttings, but bottom heat should be no higher than 16°C (60°F). The compost should be based on coarse sand with sphagnum peat, and the addition of a little crushed limestone will greatly assist rooting. Suitable young trails may also be layered into 15cm (6in) pots set into the ground where convenient around the base of the plant, and the rooting mixture in this case could be a standard commercial seeding and cutting compost, containing superphosphate and crushed limestone.

Bluebell Creeper
Sollya heterophylla (syn. *S. fusiformis*)   (Pittosporaceae)

An extremely beautiful Australian evergreen twiner with glossy, dark green leaves, ideal for the conservatory in temperate climatic zones, or for a sunny sheltered site outdoors, where it will clothe a warm wall and can be allowed to scramble freely over low shrubs. It has slender, shrubby stems which bear nodding clusters of sky-blue flowers, bell-shaped with a central pistil point, opening in midsummer and repeating freely throughout the season and well into autumn, sometimes appearing singly, sometimes in loose terminal clusters of twelve or more. The flowers, each about 1cm (½in) long, have the appearance of a harebell – the bluebell of the United States and Scotland, rather than the wild hyacinth of England and Wales – with blue, campanulate petals of star-like symmetry.

The bluebell creeper likes a deep, well-drained soil that has been enriched with leaf mould – the conditions likely to be found in dry subtropical woodland. As a plant habitat, temperate gardens more often resemble the secondary, scrubby growth which appears after virgin forests have been cleared and burnt: the humus level in the soil is impoverished, and the protection from drying winds and light frosts normally afforded by the surrounding trees is no longer available. Like many Australian plants, bluebell creeper is adaptable enough to survive these conditions, but the best results are obtained by simulating a mature dry woodland habitat. When planting the bluebell creeper outdoors, a gritty but moisture-retentive soil with thorough drainage should be provided: sedge peat and leaf mould or well-rotted manure should be incorporated with a generous surface mulch of leaf mould, and a slope of 30° or so will give better results than flat land.

At Tresco Abbey in the Isles of Scilly, bluebell creeper annually produces a great mass of tiny blue bells to festoon the ruined arches of the old abbey, keeping glorious company with the scarlet lobster-claw creeper and the blue convolvulus amidst a framework of roses and wistaria. In the cooler but still mild British region of south Wales, it clothes several house walls during the summer, but as a safety measure it is usually taken in for the winter, plants being grown for the purpose in large pots or deep wooden troughs.

Propagation can be achieved with semi-mature tips taken in late summer, or with older trails cut into convenient lengths wherever there are good leaf axil buds; preferably using material that is firm at the base while the top is still fairly soft, and avoiding shoots which carry flowers. Closely leafed trails are best, so that the basal nodes may be inserted no deeper than 2.5cm (1in), allowing the upper bud to rest flush with the surface. Cuttings should be placed singly in small pots, using a compost consisting of 2 parts coarse sand, 1 part fibrous loam and 1 part sphagnum peat, surfaced with silver sand which will trickle to the base as each cutting is inserted. Mist assists rooting, with bottom heat of 20°C (68°F), but rooted cuttings will die in too moist an atmosphere, and should be transferred to an open bench as soon as new growth appears.

Blue Dawn Flower
Blue convolvulus; perennial morning glory
Pharbitis learii   (Convolvulaceae)

An evergreen twiner from tropical and subtropical America, shrubby at the base, with variably shaped leaves and trumpet flowers of the morning glory type, intensely blue early in the morning, darkening to purple-magenta as the day lengthens. The flowers are produced in great numbers, large clusters opening successively throughout summer and autumn. In temperate zones, flowering starts once the day length has passed its summer maximum.

When one of these plants is being trained to cover a trellis, a pillar, or the greenhouse rafters, it might be remembered that, like many other vines with trumpet or bell-shaped flowers, they are to be seen at their best from below. *P. learii* likes plenty of room at the roots, and the greenhouse border should be used for planting in preference to pots; failing this, the containers should be the largest size available, filled with a fibrous loam that has been well fortified with rotted manure and leaf mould.

Other related species from tropical America include the annual *P. purpurea*, a variable plant which is to be seen with flowers sometimes scarlet, sometimes pale blue, or white; it also has a variety known as Tricolor, which has three-colour striped trumpets in red, white and blue – a type which comes true from seed and is sometimes given specific status of its own. The flowers of annual morning glories especially are very sensitive to light, and close up as night falls. Some, like *P. hederacea*, an annual which can be induced to flower whilst still very small, have been used occasionally in experiments on the response of flowering to variations of day length. The Asian species in particular are often short-day plants, needing a maximum of no more than ten hours' daylight to flower satisfactorily; longer days inhibit flowering – a familiar characteristic of autumn-flowering plants such as the chrysanthemums.

The perennial blue dawn flower grows and flowers well under glass even in quite northerly latitudes: at Dundonnell House in the north-west Highlands of Scotland, it grows so vigorously during the summer months that it occupies a large area in the conservatory, and has to be cut hard back during the winter.

For propagation, layers can be induced to strike roots quite readily, and trails of the previous year's growth should be pegged down into

10cm (4in) pots plunged below soil level in the greenhouse border, using either the tip section or a sequence of two or three loops along a trail, cutting a shallow tongue at the point to be rooted, and setting them 5cm (2in) deep in a fibrous, peaty compost. Alternatively, cuttings can be taken in early spring, using sideshoots with a heel that will need careful trimming, set in a compost consisting of 2 parts sedge peat, 1 part loam and 1 part sharp sand, incorporating a sprinkling of bonemeal. Bottom heat of 24°C (75°F) will assist rooting, with intermittent mist spray or frequent syringing with lukewarm water.

Bougainvillea
Bougainvillea spectabilis; B. glabra and others   (Nyctaginaceae)

These spectacular natives of tropical and subtropical America flourish in hot countries everywhere and have no objection to poor, dry soils. There are some differences between the requirements of the two most commonly seen species: *Bougainvillea spectabilis* climbs to some 9m (30ft) by means of the vicious thorn hooks on its stems, and appreciates in particular those climates which have a hot, wet summer and a cool, comparatively dry winter; *B. glabra*, bushier and much less fiercely armed, is hardier and more amenable to dry summers and damp winters. In the Mediterranean region both species thrive if they are given a sunny position with ample space in which to expand. Near the seashore itself, however, *B. spectabilis* and its varieties do less well, and *B. glabra* excels. The latter also does best on higher ground where the temperatures are lower, is resistant to slight frosts, and will survive the occasional cold snap such as Mediterranean countries experience from time to time, though it may lose its leaves during the winter. *B. spectabilis*, with its softer leaves and heavier, woody, thorny branches, is most at home twining high over some support; *B. glabra* is content to grow as a rounded bush and, in fact, is often used to form a colourful hedge, standing close clipping well under these circumstances.

Many of the showiest spring-flowering garden varieties derive from *B. spectabilis*, and the typical species when grown under glass in temperate zones can produce its vivid rose-purple bracts before winter has passed. *B. glabra*, with bracts of a bright magenta, has a longer flowering season in the wild, opening later than *B. spectabilis* but remaining in flower throughout summer and autumn, sometimes until the onset of winter. Red- and orange-bracted varieties and hybrids are, as a rule, the most tender: Mrs Butt is the name most often used for a natural hybrid between *B. glabra* and *B. peruviana*, correctly called *B. × buttiana*, and this is a rich red; *B. spectabilis lateritia* is a naturally occurring rusty-orange variety; McLean is a glorious orange, fading to pink; Killie Campbell has a tricolor effect of brilliant orange, pink and mauve; Orange King is bright orange and yellow – probably the same as the variety known as Mrs Louis Wathen.

Bougainvillea glabra

Bougainvilleas have no objection to comparatively low winter temperatures – dropping perhaps to 2°C (36°F) – but cold wet winters spent outdoors have the effect of limiting subsequent growth. Much depends on the conditions in early spring, for they flower on the current year's wood. During the first season following planting, a young bougainvillea should be encouraged to build up a solid woody framework before being allowed to flower. In cool temperate zones, both main species may be grown in tubs to be placed on the patio or close to a warm, sunny wall during the summer. *B. glabra* is probably the easier to treat in this way, as it takes up less room and flowers later, but even quite large plants of *B. spectabilis* may be grown in containers to be carried outside and plunged in an open garden bed for the season. During the summer months watering must be very thorough, and containers should be well crocked and provided with ample outlets to ensure a drainage system which functions perfectly. Once they have been returned indoors before the first frost, watering should be reduced until the compost is barely moist. Large plants used in this way and tied annually to trellis supports should be spur pruned when resting, leaving the main trails intact, but cutting back all the lateral shoots to their first or second bud. Any poor, weak growths should be removed at the base. This treatment will result in evenly spaced, consistently vigorous lateral shoots which will be covered with colour later in the year. Named cultivars or specimens of B. spectabilis which are to be kept compact, or confined to a small greenhouse, should be clipped and trimmed as the bracts fade to encourage new flowering wood to appear, and encourage more than one display during the season.

In a suitable climate there can be several flowerings during the year. In one beautiful garden in Barbados a fine collection of bougainvillea cultivars are grown in 30cm (12in) pots, and pruned heavily directly after every flowering so as to limit even quite old specimens to a height of about 1m (3ft). A moderately rich, peaty compost is used and this has to be topped up, or the upper layers replaced with fresh compost several times a year, while the plants are fed regularly with a high potash fertiliser containing trace elements. After pruning and shaping, the plants are watered twice daily, and foliar feed is added to their water once a week. After two weeks of saturation the water is reduced until the compost is barely moist; vegetative growth goes into check, flower buds appear, and the plants come into full bloom within a few weeks. New foliage accompanies the flowers and fertiliser is needed once again, heavy watering is recommenced, and one or two applications of foliar feed are given. After a further two weeks watering is again reduced, but this time the compost is kept moderately moist, and flowering continues for at least a month. When the bracts fade the plants are again pruned heavily and the whole process is repeated. By following this system it is normal to expect at least four peak flowering periods during the year, and the method can equally well be applied to specimens trained as small standard trees.

Bougainvilleas flower while still very small, and although feeding is essential for intensive cultivation, especially in temperate zones, unless subjected to careful control, too rich a planting compost is liable to result in lush vegetation at the expense of flowers. Large numbers of bougainvillea cultivars are grown for indoor decoration by the Brighton Corporation Parks and Gardens Department in southern England, who follow an equally strict regimen with spectacular success. During their first year the plants are grown in 13cm (4in) pots, and subsequently moved into pots of up to 30cm (12in) in which they are allowed to mature, and attain a maximum size of some 2m (7ft) high with a spread of at least 1m (3ft). As flowering ends in the autumn they are spurred back hard each year, and fed weekly during the growing season. Once the plants are two years old they are started into growth in midwinter with a temperature of 21°C (70°F), kept moist and fairly close, and sprayed frequently with lukewarm water to encourage new growth to break evenly. As foliage is produced the lateral shoots are bent over and tied in, so as to reduce the flow of sap and encourage the formation of flower buds. The glass is shaded as spring approaches, and once they are growing strongly the plants are given a weekly feed, at first containing a fairly high nitrogen content, later changed to one with a high potash ratio when the flower buds appear. As soon as the bracts show colour, the plants are moved to a lightly shaded, cooler house at about 10°C (50°F) and given full ventilation. This move causes a check in vegetative growth and encourages the bracts to mature. Watering is not diminished, however, and the weekly high potash feeds continue. Once the plants have finished flowering and shed their bracts, the longer shoots are cut back to about half their length, and a second flowering usually follows in the later summer and autumn.

In warm climates propagation of the bougainvilleas can quite easily be achieved with 15cm (6in) leafless cuttings of woody growth taken in the spring and placed in sandy compost. Prunings are often used for propagating material, and even the oldest pieces will take root if first stood in water for a few hours and then dipped into hormone rooting powder before insertion. Air layering is always effective, and conventional layers can also be taken, using pliable young shoots springing from the base of the plant. These should be given a slight twist at the point to be rooted, conveniently about 15cm (6in) from the tip, and pegged upright in a compost consisting of 2 parts sharp sand, 1 part sedge peat, and 1 part well-rotted leaf mould.

In temperate regions, and with stock plants under glass, softer cuttings with two or three leaves attached can be taken in the spring, and placed in a sandy compost with bottom heat of 21°C (70°F). After rooting the cuttings should be left in this temperature for a few days until new top growth breaks. The young plants should be grown on in an ambient temperature of 13–16°C (55–60°F) until the new growth has reached at least 20cm (8in), at which stage they should be cut hard back to encourage vigorous shoots to spring from the base.

Chilean Bellflower

Lapageria rosea   (Philesiaceae)

From Chile, and the Chilean national flower, this twining shrub is closely allied to the lilies and among the most beautiful of climbers, reaching a height of some 4m (13ft) on a wall, evergreen, with leathery, dark green leaves, and fleshy, bell-shaped flowers up to 10cm (4in) long by 5cm (2in) wide, appearing at any time of the year either singly or in clusters from the axils of the upper leaves. In the wild, flower colour varies from bright crimson through shades of carmine and pink to pure white – or sometimes bi-coloured with a marbling effect, seedlings from these forms producing very variable results.

Provided the site is well sheltered, as in a courtyard, a north-facing wall is usually the best for outdoor planting in mild temperate regions within the northern hemisphere for, like many Chilean plants, it is used to a cool mountain climate and is not injured by a little frost. Within the glasshouse it may be grown permanently in a container, but is much better planted out in beds and trained on a trellis. The beds must have thorough drainage, with an acid soil not above pH 5.5, and a good indoor growing compost might be based on 3 parts fibrous peat to 2 parts loam, with the addition of sharp sand and charcoal, watering regularly with a weak solution of potassium-rich liquid fertiliser. Outdoors or in, the Chilean bellflower needs to be kept cool and shaded during the summer, with ample water at the roots while growth is active, and a daily spray with tepid water until the flowers begin to open. With its natural habitat amid shady mountain forest, it detests long exposure to strong sunlight and needs a cool, moist, lime-free soil in shade or semi-shade, such as it might find in a sheltered garden. Only in cold districts need it be regarded as a conservatory plant. Where space is limited, the roots may be enclosed by boxing them in with slates or bricks set below the ground, allowing up to 1sq m (10sq ft) of space, for it has a tendency to spread widely when unrestricted, and often sends up shoots some distance away from the main stem.

In Californian gardens under the influence of Pacific breezes the

Chilean bellflower grows to perfection, and is often planted so as to climb a tree, where it revels in the shade of the foliage. In the British Royal Horticultural Society Gardens in Surrey it is treated as a greenhouse plant, and sheaves of its crimson flowers hang on their wiry stems from the roof ties of the temperate house there, combining with a display of jasmines and honeysuckles; but at nearby London it grows outdoors, flowering from September to December, facing west on a sheltered wall. At Tresco Abbey in the frost-free but gale-swept Isles of Scilly, it climbs a shady north-facing corner of the ruined abbey, displaying a succession of carmine bells from August to Christmas. In Hampshire on the English south coast it grows on a house wall in company with the beautiful crimson-flowered *Philesia magellanica* and a rare bi-generic hybrid between the two species — × *Philageria veitchii*. Again in England in a Somerset garden it occupies a wall facing north, where it scrambles through the branches of a tall camellia; and at Powis Castle in mid Wales, given a favoured spot on the terrace, it grows with bright blue ceanothus and tender rhododendrons such as the sweet scented Fragrantissima; nearby, on the wall of a small private house where the garden is not so sheltered, it flowers well in a container, but is taken indoors every winter.

Further north, at Dundonnell House in the Western Highlands of Scotland, it is on the verge of tenderness and grows in a conservatory in company with fruiting orange trees and the crimson coral plant, *Berberidopsis corallina*; while at the Royal Botanic Gardens in Edinburgh it grows on a south-facing wall, where it derives some shade from a silver wattle, and flowers quite well — but the best specimen to be seen at Edinburgh occupies a cool, partially shaded glasshouse. Beautiful specimens are to be seen growing outdoors in Ireland, and at Mount Usher on the slopes of the Wicklow Mountains, the pink and white forms have been planted to climb an alder together on the banks of a brook, so that the large waxy bells can hang gracefully over the water.

Rooted suckers are often obtainable from old plants, otherwise the most reliable method of propagation is by layering before growth commences in the early spring: short trails of the previous season's growth should be brought down to the ground, giving a slight twist at the point to be rooted, conveniently about 15cm (6in) from the tip; long shoots can be layered progressively in loops, pegging no more than four layers per shoot. When the stock plant is growing in a greenhouse, layers may conveniently be pegged into well-crocked seedboxes which must be at least 13cm (5in) deep, using a compost of 2 parts sharp sand to 1 part sphagnum peat, to which has been added a little superphosphate. Care should be taken to ensure that the sand does not have an alkaline reaction. Two layers can be struck in one box, the shoots bent sharply and pegged so that the base of the layers lies midway between the crock layer and the surface of the compost. By the following spring they should be ready to lift and separate, taking care not to break the brittle young roots. At this stage they should be potted individually, using a lightly fertilised but lime-free compost consisting of 1 part very fibrous loam, 1 part sphagnum peat, 1 part sharp sand and 1 part charcoal. The young plants should be kept moist and shaded, and the roots must not be allowed to become pot-bound. With outdoor stock plants, individual 15cm (6in) pots should be used initially instead of boxes, plunging them in the ground wherever the layers conveniently reach. When frost is expected outdoor layers should be given protection during the winter beneath a mound of ashes or peat; when rooting is well advanced in the spring they should be severed and repotted as above, and grown on for the summer in a cool, shaded greenhouse.

Clianthus
Parrot's bill; lobster claw; kaka's beak
Clianthus puniceus
also Glory pea; Sturt's desert pea
Clianthus formosus   (Leguminosae)

The evergreen *Clianthus puniceus* from New Zealand will grow up to 2.5m (8ft) as a sprawling, scandent shrub when provided with a warm, sun-facing wall within cool temperate areas – taller in the mildest localities and in the subtropics. The flowers which, during the summer, have bright red, long-pointed, curved-keel petals vivid against the fresh green foliage, look like the claws of a cooked lobster or, as New Zealanders say, like the beak of a native parrot. There is a white variety which lacks the exotic impact, but the two can be grown together for dramatic contrast.

Although resistant to mild frosts, prolonged cold spells will kill the top growth, and particularly severe winters will result in the death of the plant. In mild regions of Britain such as Cornwall and the south-west of Scotland, the parrot's bill succeeds outdoors on sunny walls or in a sheltered nook, and if necessary can be given some kind of thatch

62

Clianthus puniceus, the parrot's bill, grows outdoors in mild temperate regions but loses its flower buds during winter if subjected to harsh weather

Clianthus formosus, the glory pea from the Western Australian desert, needs special growing techniques which simulate its harsh native habitat

protection during the winter – but when over-sheltered in this way it is greatly subject to snail damage; snails tend to congregate in the same type of warm spot often chosen to house plants such as the parrot's bill, and the tender evergreen leaves provide a convenient filler for these creatures both before hibernation and during the early spring.

A mild western seaboard climate, whether in America or Europe, encourages healthy vegetative growth of *C. puniceus*, but successful flowering depends on the severity of frost previously experienced during the winter, for the flower buds are formed early, and if the temperature drops below – 9°C (15°F), they are apt to be killed and the flowers will fail. The vegetative growth in addition needs to be well hardened off by autumn in order to bring the buds through the winter unscathed. An answer to this problem can often be found by persuading the parrot's bill to grow through the foliage of some other plant, a mutually advantageous expedient which will often bring the flower buds safely through the worst weather. Birr Castle in Ireland not uncommonly experiences temperatures lower than – 9°C (15°F), but there on a wall, sheltered but facing south-west, *C. puniceus* flowers luxuriantly each June in intimate company with the scented *Buddleia auriculata*, tree peonies and carpenteria. Parrot's bill also thrives in the mild and windy climate of the Western Scottish Isles, where frosts are seldom severe enough to kill the buds. It will not survive saturated conditions during the winter, but rainfall on the Hebrides is only about half that of the notoriously wet western Scottish mainland; a well-drained soil and a sunny wall produce good flowering results, and the plants can grow very large. An open sandy soil is essential in damp regions, for waterlogging before or during frosty weather can be more harmful than the frost itself. Provided it is neither exposed to cold winds nor subjected to wet soil, though the flowers will be lost, parrot's bill can tolerate frosts as severe as – 14°C (6°F) without having the top growth killed.

While south-west England is often mild enough to grow *C. puniceus* satisfactorily, the drier south-east as a rule is not. In a certain Kentish garden a large and very vigorous specimen occupies a sheltered position in a courtyard, the wall facing east but protected from the bitter continental winds. In this situation it covers 1sq m (10sq ft) or so of wall each year, but has never yet succeeded in flowering, despite the presence of flower buds in the early winter. At the Royal Horticultural Society's Gardens at Wisley, where plants must give of their best, parrot's bill used to occupy a glasshouse kept at a winter minimum of 4°C (40°F). There also it was given an easterly aspect so that the leaves at least could feel the morning sun, and it was one of the earliest climbers to flower each spring, with heavy clusters of blood red claws hanging below the dense screen of foliage.

Also under glass at Wisley is the early spring-flowering *C. formosus*, the glory pea from the Western Australian desert. An evergreen, semi-trailing plant with silky-fine silvery foliage, it rarely grows taller than 1m (3ft), and bears clusters of about six brilliant red flowers each with a raised purple-black centre, of similar overall shape to the parrot's bill flowers, but arranged radially around the stem in a colourful whorl. In the Botanic Gardens at Perth, Western Australia, the glory pea is truly at home in an area where the relative humidity in the summer is 55 per cent, and in the winter 69 per cent, and where the shade temperature reaches over 32°C (90°F) on at least thirty days of the year. During the winter, ground frost occurs only rarely and very lightly. The area is notably windy, a factor which may be significant when difficulties are encountered in colder climates, where the glory pea is grown as a greenhouse plant, for it is plainly conditioned to a very free circulation of air. This fact is reflected by its performance in mild and very windy places like the Isles of Scilly, where both *C. formosus* and *C. puniceus* can flourish outdoors throughout the year.

C. puniceus is relatively easy to propagate and takes fairly readily from cuttings. As soon as the flowers start to fade, nodal cuttings should be selected from amongst the topmost shoots, their length depending on the distance between nodes, but usually around 8cm (3in). These should be placed singly in small paper pots, using a compost of 2 parts loam, 1 part sphagnum peat and 1 part sharp sand, given mild bottom heat of around 18°C (65°F) and covered with very light-gauge polythene sheet that must be removed as soon as growth starts. Alternatively, 10cm (4in) sideshoots can be taken about a month later, using firm wood pulled off with a heel, and these may be set in an unheated closed frame, using pure sand as a rooting medium. During warm weather careful attention is needed to keep the sand uniformly moist. As soon as top-growth starts after four or five weeks, the rooted cuttings should be potted immediately, using the standard compost of 2 parts loam, 1 part sphagnum peat and 1 part sharp sand, and the pots set in an open, airy place.

C. formosus will not grow easily from cuttings. The seed germinates readily, but there is a marked tendency for the young seedlings to die before they reach the flowering stage. The conventional solution to this problem has been to graft the young glory pea seedlings on seedlings of the common senna, *Colutea arborescens*, and these then grow well; but the operation is extremely delicate and fiddling, and many gardeners, both amateur and professional, are quickly discouraged from attempting to rear them by this method. The major cause of death in glory pea seedlings seems to be a type of collar rot infection which strikes them just below the soil surface; in their desert home they normally escape this trouble. There in the climate of Western Australia the seeds germinate as a rule following a sudden storm – often the last rain they are to experience for some time. The vigorous taproots follow the disappearing water down into the sandy soil, enabling moisture to be drawn from the deeper layers, while the root collar remains dry. This situation can be simulated in the greenhouse by sowing the seeds barely covered in a compost consisting of 2 parts sharp sand, 1 part loam and 1 part sphagnum peat, using deep individual pots set over bottom heat of 24°C (75°F), but left uncovered to allow free circulation of air. The compost should be well soaked initially, but as soon as germination takes place – which normally occurs before the compost has had a chance to dry out – no further water should be given from the top, irrigating only via the base of the pots. Once the young glory pea plants have reached the flowering stage, there should be no further danger from collar rot, and they may be grown on in the normal way.

Cobaea
Purple bells; cup and saucer vine; cathedral bells
Cobaea scandens   (Polemoniaceae)

An evergreen from the drier regions of tropical America and north to Mexico, climbing with the aid of tendrils to some 6m (20ft). The leaves are rubbery and smooth, of a fairly deep green, and the

(*Above left*) *Aloe arborescens* is one of the easiest aloes to grow outside its native environment. It makes a spiny mass several paces across. (*Above right*) *Lithops* spp. Living stone plants have evolved a shape and life style which enables them to survive in harsh desert regions, but they are easy to grow in a cool greenhouse. (*Below left*) *Cestrum purpureum* is a rambling evergreen that will survive mild temperate winters outdoors, but it grows better under glass where frost can be excluded. (*Below right*) An indoor display of protea blooms. The globular flowers surrounded by colourful bracts last a long time after cutting and are often used in dried flower arrangements

strikingly beautiful flowers, nodding on 18cm (7in) stalks, measure 5cm (2in) long by 4cm (1½in) across – pale green at first, changing to greenish mauve, followed by rich purple, with flowers of all three colours on display simultaneously. As they mature they remain green within the throat, and when seen from below this colour contrasts charmingly with the purple of the full-blown bells. When the flower withers the campanulate cup falls from the saucer, but this saucer-shaped calyx remains, pale green and highly decorative. Flowers appear continuously throughout the summer months, and are succeeded by tiny cucumber-like fruits. There is a white-flowered form, *C. scandens flore-albo*, and also a variegated form, Variegata, which has white-blotched leaves.

When grown in a greenhouse, cobaea is one of the fastest-growing plants, producing 3.5m (12ft) trails in one season, but it may be cut back without harm and treated as a shrubby perennial – or even an annual, as it grows quickly from seed. In the greenhouse or conservatory it is best treated as a perennial, and planted out in a fairly rich soil with a bed of its own, where it fares better than when the roots are restrained in a pot. Its watering requirements are modest, and it prefers a fairly dry compost in the winter. When grown under glass as a permanent climber it quickly covers walls or rafters, and the long shoots can be pruned back in the autumn as convenient. The variegated form is far less vigorous.

Young specimens should be tied to a support until the tendrils develop and the shoots are able to cling naturally. As a climbing half-hardy annual planted outdoors cobaea is capable of covering a large area in the course of a growing season, and in sheltered gardens within temperate climatic zones it will flower throughout the late summer and autumn until the first frost kills it. In town gardens especially it has proved valuable to cover the roof of a summer house or porch, and it will outclass the Russian vine *Polygonum baldschuanicum* in forming a rapidly grown, temporary cover while the permanent plants are slowly growing into place. On sunny walls, it can act as a very useful support and backing for other climbers such as the passion flowers – a use to which it has been put very effectively on the sunny terrace of Sternfield House at Saxmundham in East Suffolk.

Petrea volubilis, the purple wreath from tropical America and the West Indies, will tolerate fairly low temperatures, but needs protection from frost

Ripe seed germinates readily when fresh, retaining its viability only for a few months. Sowing should be carried out very early in the spring over 16°C (60°F) bottom heat, using a standard seed-sowing compost. Seedlings should be potted up as soon as they are large enough to handle, into a compost of 3 parts fibrous loam to 1 part sphagnum peat and 1 part coarse sand, feeding regularly with strong liquid fertiliser if they are to be raised in containers. Moved evenly into larger sizes until they occupy a 30cm (12in) pot, specimens will be in bloom within four months and, by the autumn, will have clothed perhaps 15sq m (160sq ft) or more of wall space. If seed is not available – and necessarily in the case of *C. s.* Variegata – 8–10cm (3–4in) cuttings may be taken of the young shoot tips as they grow from pruned stock plants in the spring, and set in a sandy compost, using bottom heat of 18°C (65°F).

Flame Nasturtium
Scotch creeper; Scotch flame flower
Tropaeolum speciosum   (Tropaeolaceae)

A succession of brilliant scarlet flowers during the summer months, set amid fresh green foliage, typify this perennial creeper from the cool, moist mountain climate of Chile. Its alternative British names suggest the conditions which suit it best, but although most at home in those climatic zones within the northern hemisphere that equate to Chile – the Pacific coastal regions of Washington and Oregon; the Atlantic Gulf Stream climate of Scotland, Ireland and Wales – *T. speciosum* can be grown outdoors anywhere within the temperate regions, given shade, summer moisture, and moderate winter temperatures. It will happily cover a bleak, sunless wall where the root area and at least the lower part of the stem is permanently shaded, provided the soil is moist and lime-free. Scotch creeper is able to survive in limy soils, but truly healthy growth within limestone or chalk areas is quite exceptional, and it prefers fairly acid, peaty conditions. It is not a plant for the specially prepared peat bed, however, for once established in such a choice site it can prove almost ineradicable, and in several parts of the wet north and west it has become something of a weed – albeit a beautiful one. In districts where the rainfall is less than 90cm (35in) per annum, and the winters are mild, it will grow demurely and is very useful to provide extra summer colour for shrubs; deciduous azaleas, for example, are greatly

A quick-growing tropical climber, *Cobaea scandens* may be treated as an annual in cool climates, when it will produce a succession of large purple bells until the first frost kills it

Tropaeolum speciosum, the flame nasturtium or Scotch creeper, is a Chilean herbaceous perennial which can provide a splash of bright scarlet in damp northern climates

enhanced during the dull period between spring flowering and the onset of their autumn leaf colours, if a Scotch creeper is allowed to climb over them.

Both in Scotland and in Ireland *T. speciosum* has frequently been planted so as to intermingle with dark yew hedges. At Castlewellan near the Mountains of Mourne in County Down, it rampages at will through the grounds, adding spectacular colour to the hedges, shrubs and conifers; but near the south coast of England in comparatively dry Sussex it romps through the rhododendrons almost as freely, without becoming a nuisance. In the wild garden at Wisley in Surrey it enjoys a cool root-run as a woodland plant, and climbs freely amongst the azaleas and other shrubs, while in Cornwall — perhaps the mildest region of mainland Britain — it features in many a garden and has been seen growing through a large specimen of the hybrid *Corokia* × *virgata*, where its late scarlet flowers contrast very effectively with the amber berries. In Devon, it has been placed to cover a north-facing wall where it makes a magnificent red and green backdrop to a group of rhododendrons and magnolias; and across the Bristol Channel in a Welsh garden it scrambles over a tree stump with *Aconitum volubile*, the two plants intertwined in a profusion of scarlet and violet flowers.

Many gardeners experience a great deal of difficulty in first establishing the Scotch creeper, for the fleshy rhizomes resent the disturbance of being divided and moved. Young seedlings take fairly easily provided they are planted in cool, moist, peaty, lime-free soil and given the support of twigs or pea-sticks during their early stages of growth, but the stick-like rhizomes are very brittle and, if broken, tend to remain dormant for a whole season, though they will sprout again the following year. Roots bought from nurseries in the spring are almost invariably dried out, and this also has the effect of holding growth in check until new annual buds can be produced. The safest time to move an established root outdoors is during active growth, replanting it immediately at least 23cm (9in) deep. Closely related annuals are more easily raised from seed: *T. canariensis*, the canary creeper, with lemon yellow flowers, will germinate rapidly and cover quite a large area within a short time; both this and the popular annual sprawler *T. majus*, the common nasturtium — so readily raised from seed sown in the open during the spring — might advantageously be used to prepare the way for *T. speciosum*, temporarily covering the ground or a trellis, providing the moist shade at root level appreciated by the perennial creeper, and incidentally keeping weeds down so that the new plants need not be disturbed by weeding as they are becoming established.

Scotch creeper seed should be sown under glass just as the outdoor buds start to open, using a compost of 3 parts lime-free fibrous loam, 2 parts sphagnum peat and 1 part coarse sand, with the addition of a sprinkling of superphosphate, and setting the seed box over bottom heat of 16°C (60°F). The same mixture could be used for cuttings if seed is not available: tips and nodal sections of the slender, wiry shoots, about 6cm (2½in) long, taken as growth starts in the spring, should be set fairly close together in a deep seed box over bottom heat of 18°C (65°F), and kept lightly shaded. Many of them will have callused and formed rhizomes by the autumn. The rooted cuttings should be kept cool and sparingly moist during the winter, giving only enough water to prevent the desiccation of the fine roots, and potted up in the late spring, using a lime-free compost containing a high proportion of sphagnum peat.

Magnolia
Climbing magnolia; laurel-leaved tulip tree; bull bay; laurel magnolia
Magnolia grandiflora (Magnoliaceae)

This evergreen magnolia, thriving as it does over a wide range of soils and climates, must rank as one of the world's most widely planted exotic broadleaf trees. Fine specimens are to be found in every continent, in conditions ranging from the Pacific breezes of Seattle, where the variety Goliath is magnificent, to the subtropical mountain climate of Kathmandu, where mature trees grow in the royal palace gardens.

The creamy white flowers measure up to 25cm (10in) across, with thick waxy petals and contrasting purplish-red stamens, exuding a fruity, lemon-like fragrance which is particularly noticeable towards evening. Usually only a few blooms open at any one time, but during a warm season they appear successively throughout summer and early autumn. Its thick, leathery, glossy green leaves have a reddish-brown fur on their undersides which makes young trees in particular look russet when the foliage is stirred by a breeze. After a long hot summer, large cone-like fruits appear, displayed in a ruff of dark green leaves, and these split as they ripen to reveal vivid orange seeds. In regions where the summers are cool and dull and the flowers correspondingly late to make their appearance, seeds are rarely

73

produced, although blooms may continue to open until well into autumn if the weather stays mild. In its native south-eastern USA the climbing magnolia flowers from June or July onwards, allowing ample time for fruit ripening during the long, hot summers.

Quite strongly lime-tolerant, *M. grandiflora* is one of the few magnolias to thrive in a chalky soil, but in such places particularly spring droughts can be dangerous, and newly planted young specimens should be given a surface mulch of leaf mould or peat; the roots should be kept well watered and, in dry weather, the foliage should be sprayed in the evening. A west-facing wall provides a very satisfactory site, as a rule, in the British Isles. At Kew Gardens in London the large wall-trained specimens of climbing magnolia, particularly the varieties Goliath and Exmouth, produce better flowers than the nearby free-standing trees. Exmouth has very large, very fragrant flowers, and those of Goliath are even larger, though rather less fragrant. Both varieties flower at an early age, but Exmouth in particular starts to bear when no more than 1m (3–4ft) high.

Grown in the open, *M. grandiflora* can be smashed by heavy snow, as happened at Windsor Great Park in the winter of 1963; Exmouth usually escapes this type of damage because of its stiff, erect habit. At Hodnet Hall in Shropshire, England, where many trees and shrubs were damaged by a season of severe spring frosts and bitterly cold winds, climbing magnolias survived unscathed, but during the same year many specimens were badly damaged at Bodnant, in north Wales, and several died. Aspect can have a dramatic effect on growth and flowering: at Hodnet Hall there are two very old wall-trained specimens, one facing east – a stunted plant which never flowers, and one facing west – tall, vigorous, and a superb flowerer. A few miles away on a high south-facing wall, a twenty-year-old Goliath flowers annually in intimate association with a mauve wistaria, a claret-leaved vine, and a red climbing rose.

The climbing magnolia can act as a shelter for tender climbers, as it does on the wall of Knightshayes Court in the English West Country, where it nurses the sweetly scented white *Trachelospermum jasminoides*, white and yellow banksian roses, and the pink climbing gazania. In County Kildare, eastern Ireland, a fine specimen on a high south-facing wall has been interplanted with a tender blue ceanothus, and at Powis Castle in mid Wales, it is to be seen at home on the sheltered Orangery Terrace, blending with the brilliant orange-flowered pomegranate and the blue passion vine.

The evergreen climbing *Magnolia grandiflora* can bring an exotic touch to dull northern gardens

Plants grown from cuttings can be expected to flower much earlier than seedlings, which may not bloom for twenty years or more; the varieties, though early flowering will not, of course, come true from seed. The ideal season to take cuttings of *M. grandiflora* and its varieties corresponds with the appearance of the flowers – whether they open in June as they do in the southern USA, or in September as they do in Britain – for this timing ensures the optimum stage of ripeness. Half-mature sideshoots should be used with a carefully trimmed heel, the length of cutting depending entirely on the vigour of growth. They should be set singly in small propagating pots, using a compost consisting of 2 parts loam, 1 part sphagnum peat and 1 part coarse sand, embodying a sprinkling of superphosphate, and surfaced with a little sharp sand which is allowed to fall around the base as the cuttings are inserted. After watering well in, the pots should be placed over 16°C (60°F) bottom heat and kept well shaded during sunny weather. Intermittent mist spray is ideal, but if this is not available they should be syringed periodically with tepid water. Rooting should have taken place within two months, and the rooted cuttings should be kept in an open frame until they are growing strongly.

Mandevilla
Chilean jasmine
Mandevilla suaveolens and others   (Apocynaceae)

This evergreen climber from the Argentine produces sweetly fragrant waxy white flowers in the summer, elegantly funnel-shaped with curved-back petals, each 5cm (2in) across the mouth, usually in clusters of four suspended from the leaf axils. A slender, sun-loving twiner for trellis, pillar or pergola, capable of reaching 4.5m (15ft), with narrowly heart-shaped leaves distinctively tufted with white down between the under-surface veins. It will not stand severe frost or cold winds unprotected, but is luxuriant of growth in a fertile soil where frosts are mild; if planted out in temperate regions such as the milder British counties it needs the protection of a warm sheltered wall and a well-drained soil. It is perfectly happy in Cornwall when sited in a sunny niche, though even there it tends to lose its leaves during the winter, and it has been grown successfully outdoors in sheltered gardens in the English Midlands, where the top growth is annually killed back by frost but the roots survive, if they are given a protective covering of ashes and bracken. In County Tipperary in south-central Ireland it grows well on a wall in company with the tender *Clematis paniculata* Lobata, and the climbing gazania *Mutisia decurrens*, which flowers with the mandevilla in a blaze of orange during August, but here, too, they are given winter protection. In County Kildare in eastern Ireland it survives outdoors in a small, sunny courtyard which it shares with the deep blue spring-flowering *Ceanothus impressus*, *Rosa xanthina* Canary Bird, and the scarlet trumpet honeysuckle, *Lonicera × brownii*.

Unlike much of the USA, summers in north-western Europe tend to be too cool and short to ripen subtropical fruit properly, and in Britain the Chilean jasmine needs greenhouse conditions to produce seed pods regularly – which it does in the temperate house at Wisley Gardens, where a continuous succession of flowers in late summer and autumn are followed by the pencil-thick pods which grow in pairs, up to 30cm (1ft) long, often joined distinctively at their tips as well as at the base, but free in the centre. As a greenhouse plant, it rarely

The white flowered Chilean jasmine, *Mandevilla suaveolens*, is an evergreen twiner which needs the protection of a warm, sheltered wall

succeeds permanently in a pot, but needs planting into a border of fertile, turfy loam which contains a proportion of roughly fibrous peat and sharp sand.

M. suaveolens is the hardiest of the genus; *M. boliviensis* needs slightly warmer conditions, but makes an attractive conservatory plant reaching no more than 3m (10ft), with bunches of glistening white, golden-throated tubular flowers hanging below glossy foliage; *M. mysorensis* from India also needs subtropical conditions, and mature specimens are quite spectacular when their foliage is hidden beneath masses of red and yellow flowers, but they start to flower only when several years old.

Propagation of *M. suaveolens* may be achieved using cuttings of small, stiff, semi-mature sideshoots some 8cm (3in) long, taken in midsummer. They should preferably be firm at the base and cut flat below the basal bud. Small individual pots are convenient, with a compost consisting of 3 parts sharp sand to 1 part sphagnum peat, and after watering well they are placed in a shaded frame and given 27°C (80°F) of bottom heat. As soon as growth starts they should be transferred to a closed frame, admitting air gradually and potting on once they are growing vigorously, into a well-fertilised, fibrous, turfy loam to which has been added a high proportion of sharp sand and a small quantity of sedge peat. Seed, if available, can be sown very early in the spring, using a compost of 2 parts finely sifted sphagnum peat to 1 part sharp sand, with a pinch of superphosphate and ground limestone. The seed should be covered only very lightly, watered gently, and the box placed in an airy greenhouse over bottom heat of 20°C (68°F). When germination is apparent the box should be kept lightly shaded, potting the seedlings on as soon as they are large enough to handle, using a standard commercial potting compost.

Mistletoe

Viscum album; Phoradendron flavescens and others   (Loranthaceae)

Christmas is the time berried mistletoe sprigs are most in evidence – certainly in England, where not all those offered for sale in the market place are native-grown, many having been imported from northern France. A relic of mysterious druidic rites, this ornamental parasitic 'climber' has a fascination beyond its more obvious cash value, and many attempts, usually unsuccessful, have been made to establish it on garden and orchard trees.

In America the place of the British and European pearl-white berried *Viscum album* is taken by the less hardy but rather similar *Phoradendron flavescens*, with amber-white berries. This is found mainly on tupelo and red maple trees over a fairly wide range from central New Jersey to Missouri, and south to Florida, Texas and New Mexico. Elsewhere in the USA, the more vigorous western mistletoe *P. flavescens macrophyllum* forms large clumps, as a rule on poplars and willows from Texas to Central California; while along the Pacific coastal region from Southern California northwards as far as Oregon, another mistletoe species, *P. villosum*, is to be found chiefly parasitising oaks.

In the case of the European *Viscum album*, apple trees are probably the commonest host, and in England mistletoe is also seen very frequently on poplars, willows and limes; it also grows on hawthorns, maples and mountain ash, and occasionally on oak and pear. Other races of the same mistletoe species on the European continent are found, in one case, only on conifers, particularly the Scots pine; or on various broadleaf trees, but with distinct strains each specialising in a particular species. In Middle Eastern forests, *Parrotia persica* is often used as a host.

One of the greatest obstacles to success in attempting to grow mistletoe under cultivation is the disparity between the regular season of harvesting, and the seasonal ripening of the berries, for those obtainable at Christmas time are immature, and can scarcely be expected to germinate when saved from sprays cut during the winter. Berries of *V. album* ripen in the spring and, as a rule, will germinate only at that season. They are not particularly attractive to birds, and often survive the winter without being eaten, but should there be a particularly severe spell of weather the mistletoe bushes are liable to become stripped of their fruit very quickly. Cut sprays begin to deteriorate fairly rapidly and, as the seeds will not continue to mature after gathering, germination success depends on their remaining untouched on the bushes until spring. A reliable way to ensure that ripe berries will be available for sowing at the correct time is to cover the whole bush with a large muslin bag soon after the berries have been formed, securing it around the neck of the plant at the point where it grows from the host branch. Some selection is necessary as mistletoe may be either male or female, with perhaps twice as many females as males, and nothing will be achieved by covering a male plant.

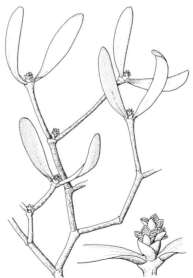

Viscum album

Mistletoe berries burst if they are picked individually, so in harvesting them, unless they are to be applied immediately, it is necessary to cut the spray with berries attached. In the spring, cut sprays soon lose their moisture content and the ripe berries start to shrivel and spoil, and if they are not to be used immediately they are best hung in a cool, airy place. Even then, particularly if the weather is hot, deterioration sets in after only a few hours. Once the berry skin has been broken, the thick viscid pulp sets hard after a few days, and it then becomes difficult to position the seeds effectively. Smeared on the host branch at the right stage, the pulp adheres to the bark and holds the solitary seed in place, and it seems to make no difference whether the seed is placed above or beneath a branch. However, when the berry is broken open and smeared immediately on the twig, particularly if the bark of the host branch is smooth, the pulp still retains enough fluidity to allow the seed to roll downwards and droop out of position; it is sometimes more convenient to smear the berries initially on a piece of wood and wait until the pulp becomes tacky, the hardening process beginning as soon as it is exposed to the air. Once the seed is firmly stuck to the branch it is relatively safe from marauders. If the seed skin remains adhering to the bark, however, birds are liable to see this and pull it off, taking the seed away with it, and for this reason the skin is best removed at the outset and discarded. At this stage an excess of pulp can act as an effective bird-lime, and care should be taken to see that no small birds become trapped.

The bark of the host branch should be left intact at the point of application, and not scraped or cut in any way; young, thin bark is more readily penetrated by the emerging delicate radicles, and the best results involve the use of first-year shoots on the host plant, at some distance from the main stem. *V. album* takes four or five years to become established and commence vigorous growth, by which time of course the host branch will have thickened proportionately.

Seeds which are brown or soft should be discarded; viable seeds are flat and heart-shaped, and green or yellowish in colour. As they germinate and the root-like radicles emerge to penetrate their host under cover of the hardened pulp, adequate moisture seems essential to success, and daily syringing with water is advisable at this stage. Berries which have been left on their parent bushes after late spring may begin to germinate *in situ*, sometimes sending their radicles into the existing host, or occasionally into the parent tissue as secondary parasites. If smeared on the bark at this stage, the developing radicles

By using the correct method, *Viscum album*, mistletoe, a relic of mysterious Druidic rites, can be established successfully in the garden

Mutisia oligodon. This South American climbing gazania with pale pink flowers retains a shrubby stature and is easier to grow than others of its genus

are certain to be damaged; harvesting therefore should not be too long delayed.

After a successful take, a young mistletoe plant will often remain apparently dormant for several years, for external growth cannot be produced until the parasitic roots or haustoria have become firmly established, forming a characteristic swelling beneath the host's bark. This period of development may be shortened by transplanting a piece of host tree bark, taken from the base of an existing mistletoe bush, and patch budding this on to an intended host of the same species, sealing the introduced patch into place with adhesive plastic tape or low-melting-point wax.

Some of the world's mistletoes have brightly coloured berries. One particularly ornamental species from the Mediterranean area, to be found in the hills of southern Spain, in Morocco and the Lebanon, is the red-berried *Viscum cruciatum*, which parasitises mainly olive trees, and has also been recorded growing on ivy. Tropical mistletoes develop far more quickly and vigorously than any of the viscum or phoradendron species, a few producing external roots which grow along the host branches and penetrate the bark in several places. The bright orange-berried *Phthirusa pyrifolia* from tropical America spreads rapidly in this manner, and has been grown successfully in the tropical greenhouses of the University of British Columbia in Vancouver, using host plants of oleander, codiaeum and citrus. The method of applying seeds to the bark is identical: germination similarly takes place within a few days; but in contrast with the temperate mistletoes the resultant phthirusa plants are fully established, flowering and berrying freely a mere year later.

Mutisia
Climbing gazania
Mutisia decurrens; M. clematis; M. oligodon (Compositae)

These magnificent South American climbers are not very easy to grow well, and need favourable conditions. A wall facing the sun is often best in temperate climatic zones, and they enjoy the company of other plants with which they can mingle.

The mutisias climb by means of tendrils at the tips of their lanceolate leaves, and have colourful, daisy-like, spidery petalled flowers like a huge herbaceous gazania. *Mutisia decurrens* from Chile usually climbs to about 3m (10ft), and produces perfect 13cm (5in) orange daisies during high summer, prominently displayed above delicately tiny leaves. As many as three hundred flowers have been counted on one plant during the course of a single season. It does not usually produce seed in Britain or the more northerly United States, though it flowers outdoors and has been known to withstand temperatures as low as −10°C (14°F).

Less hardy but climbing vigorously to 9m (30ft) under glass, or outdoors in climates similar to its native Ecuador, *M. clematis* has vivid orange-scarlet pendulous flowers produced from spring to autumn, and leaflets which are woolly when young, becoming smooth as they mature, the central leaf stalk terminating in the typical long tendril. So vigorous is the species under glass that it needs heavy cutting back each year.

M. oligodon is the easiest of the genus to establish but the least vigorous, often retaining the stature of a shrub and not usually climbing above 2m (6½ft). It produces 10cm (4in) flowers of a clear satiny pink, and climbs with the aid of long tendrils curling from the ends of the leaves, which are a shiny dark green on the upper surface, and woolly grey beneath. On the wall of Knightshayes Court in the West Country of England it flowers well each year and survives the winter in company with banksian roses and the climbing magnolia; but in Wisley Gardens in Surrey it is considered tender enough to occupy the half-hardy house, where it flowers profusely from August to September; it grows outdoors in County Tipperary, Ireland, where it enjoys the protection of a sunny wall, but is covered up annually as a winter precaution; in Scotland at Crathes Castle in the Dee Valley near Aberdeen, it grows outdoors and flowers well, only requiring winter protection when it outgrows the top of its sheltering wall.

M. decurrens similarly needs a sheltered spot when planted outdoors in the British Isles: at Bagshot in Surrey it flourishes on the sunny wall of an enclosed courtyard − perhaps the ideal site for it; near the east coast of Ireland in County Wicklow it luxuriates amid a beech hedge, so that the beech startlingly seems to bear enormous orange flowers in the late summer; several gardens in the English West Country also have the orange climbing gazania intermingling very effectively with shrubs and hedges of different kinds, or, in sheltered spots near the sea, sometimes trained along a fence to make a hedge of its own.

As it matures, *M. decurrens* begins to spread below ground and sends up new shoots several feet away from the main stem, and these suckers may be dug out with part of the root system attached and replanted. It is not an easy plant to propagate with soft cuttings under mist; if no suckers are available, older material should be used, cutting the trails into convenient lengths at the nodes, and placing these cuttings in comparatively cool and dry conditions within a cold frame.

To propagate the other climbing gazanias, in the case of plants grown outdoors, cuttings should be taken in late summer; from greenhouse plants, in the spring. Half-mature sideshoots about 8cm (3in) long and firming towards the base, are taken with a slight heel. They are best set around the rim of a 12cm (4½in) pot, using compost consisting of 2 parts sharp sand, 1 part sphagnum peat and 1 part loam, surfaced with 1cm (½in) of fine silver sand before inserting the cuttings, and the pot placed in a shaded frame given bottom heat of 20°C (68°F). As soon as new growth appears, transfer the container to a closed frame, potting the rooted cuttings singly when they are growing strongly. The atmosphere should be kept close, and the plants hardened off gradually before being introduced to the cool greenhouse.

If seed is available, it should be sown in boxes during late winter or very early spring, the compost consisting of 3 parts sphagnum peat to 1 part coarse sand, and embodying a little superphosphate of lime. The seeds should be covered lightly with fine peat and watered very gently, placing the boxes over 16°C (60°F) of bottom heat. When germination has taken place the seedlings should be kept shaded, and potted on as soon as they are large enough to be handled easily.

Pandorea
Australian bower plant
Pandorea jasminoides
also Wonga-wonga vine
Pandorea pandorana   (Bignoniaceae)

The Australian bower plant is an evergreen twiner climbing without the aid of tendrils to 9m (30ft) or more, producing great masses of trumpet-shaped flowers, each 5cm (2in) long by 4cm (1½in) wide, white on the outside, tinged with pink and crimson in the throat. This internal colour is often very pale, and there is a pure white form of the species known as *P. jasminoides* Alba. It is a frost-tender plant which needs a rich soil and plenty of water during the growing season, and a sunny position with greenhouse protection is necessary when grown in temperate climates.

The closely related wonga-wonga vine has very similar requirements. Its foliage, when mature, is also similar, although the young leaves are comparatively fine and ferny; the flowers are smaller, each measuring about 3cm (1¼in) long by 1cm (½in) wide, creamy white on the outside, with brownish crimson markings in the throat.

To suit either species, the soil in the greenhouse border should be of the woodland type: basically a sandy loam full of fibre, with additional humus in the form of sedge peat and well-rotted garden compost or leaf mould. The pandoreas are seldom completely happy when planted in containers. Like many woodland climbers, they appreciate shade at their roots and on the lower part of their stems, but they need to feel the warm sunshine on their upper foliage, and their flowering performance depends on light rather than warmth. Watering during the growing season must be copious, and the beds accordingly need to be very well drained. During the winter, however, the compost should be no more than moist. A summer growing temperature of 21–24°C (70–75°F) is adequate, and the winter temperature should not be allowed to fall below 7°C (45°F).

Cuttings can be taken in the early summer, using either half-ripe sideshoots, or the topmost third of a trail, cut between nodes into lengths of about 5cm (2in), inserted so that the node with its leaves attached is resting on the soil surface. Individual small pots are convenient, using a compost consisting of 1 part sharp sand and 1 part sphagnum peat. The propagating frame should receive as much light as possible, provided the leaves do not become scorched, and a mist spray or syringe can be used during bright sunshine, allowing the foliage time to dry before evening. Bottom heat need be no higher than 16°C (60°F), but the frame should be kept closely shut while rooting is taking place. The rooted cuttings should be hardened gradually to the open greenhouse before potting them on into a moderately fertilised sandy compost which contains some fibrous peat and leaf mould.

Passion Flower
Passion vine; passion fruit; granadilla
Passiflora caerulea and others   (Passifloraceae)

The blue passion flower, *Passiflora caerulea* from South America, can make a rampantly vigorous, dense mass of tangled stem comparable to a well-grown *Clematis montana*, evergreen in mild areas and able to survive the less severe temperate winters outdoors. The slightly fragrant mauve-blue fringed flowers, 8–10cm (3–4in) across, appear successively throughout the summer and autumn until the first frost. The leaves are apt to shrivel during cold weather and, when planted outdoors in the northern hemisphere, a sunny, sheltered, south-facing wall often provides an ideal site. The orange fruits of *P. caerulea* and its varieties are very ornamental when they appear in the autumn, but within temperate zones these ripen only after a long, hot summer.

In northern Britain the blue passion flower appreciates the protection of a greenhouse, and though it has been seen growing and flowering in the lee of a wall as far north as Leeds, it is plainly unhappy there. At the University of London's Wye College, in Kent, following a very cold winter which killed many plants, *P. caerulea* lost all its top growth on a south-facing wall, but recovered during the spring. Micro-climate is often of more significance than locality; on the sheltered Orangery Terrace at Powis Castle in mid Wales, the blue passion flower grows luxuriantly year after year above and behind the flowering pomegranates and magnolias.

In greenhouse borders, the depth of soil is best restricted to about 30cm (1ft) so that vegetative growth will not become too luxuriant to the exclusion of flowers; but it needs perfect drainage and plenty of water during the growing season. If the soil is over-rich, foliage again will benefit at the expense of the flowers; a turfy loam is best, with an admixture of peat and sharp sand. Weak shoots should be cut away to prevent overcrowding, allowing the best trails to arch gracefully. The blue passion flower is sometimes grown as a houseplant in an 18cm (7in) pot, using a moderately fertilised compost which might consist of 2 parts fibrous loam, 1 part sphagnum peat, and 1 part sharp sand. Growth should be kept well trimmed, and the plants given a definite resting season by allowing the compost to dry out slightly during the winter. Beautiful but very vigorous varieties of *P. caerulea* include the mauve Grandiflora and the ivory-flowered Constance Elliott, both of which will cover a large greenhouse roof area, and need plenty of room to ramble, growing sometimes too rampantly in the standard peaty loam.

The granadilla, *P. edulis*, is cultivated in many countries for its fruit. It has 12cm (4½in) white flowers, tinged with lilac and purple, followed by oval fruits the size of an egg, green at first, becoming dull black or dark purple. It is rather more tender than the blue passion flower, but is very vigorous, and hardy to slight frost. Young plants

When planted outdoors in the northern hemisphere, *Passiflora caerulea*, the blue passion flower, appreciates a sunny, sheltered, south-facing wall

Passiflora quadrangularis, the giant granadilla, is a lovely, rampant climber with flowers of blue, claret and pink

especially need the protection of a warm, sheltered position, and benefit from a generous mulch of well-rotted manure. The very ornamental foliage makes a thick screen of rich, shining green, and this can be cut hard back in the winter without ill effect. A light soil is best for *P. edulis*, mildly enriched with a modest addition of manure and garden compost, but a mulch is always beneficial. Young plants grow quickly and bear few fruits the first year, a large crop appearing the second year; but it is a short-lived vine, lasting only about eight years. (For fruit production, *see* page 123.)

Hybrid passion flowers include: Allardii, with white, pink and blue flowers up to 12cm (4½in) across, a strong grower surviving outdoors in the milder temperate regions; Exoniensis, suitable for the conservatory, with pendulous pink and white flowers up to 13cm (5in) across; and John Innes, with 13cm (5in) nodding cup-shaped flowers of purple and white, ornamental and free flowering under glass within temperate zones.

The species *P. antioquiensis*, which survives outdoors only in very mild areas, has pendulous flowers of vivid crimson and violet, about 13cm (5in) across, appearing in the late summer and autumn. It needs planting in comparatively rich but shallow soil inside a large greenhouse where the temperature can be kept above freezing during the winter. Other species include: *P. jamesonii*, with coral-red flowers; *P. manicata*, which is less vigorous, and has scarlet flowers with a blue central fringe; *P. molissima*, the banana passion fruit, an evergreen with pink flowers, climbing to 6m (20ft) or more and producing edible 7cm (3in) yellow bananadillas; *P. umbilicata*, with small violet flowers, one of the hardiest and most vigorous species which survives outdoors in southern England; and finally the giant granadilla *P. quadrangularis*, grown under glass at Wisley Gardens, a lovely rampant climber for the cool house, producing 13cm (5in) pendulous flowers of blue, claret and pink.

Seed of passion flower species should be sown in the earliest spring, using a standard seed sowing compost with bottom heat of 24°C (75°F). *P. edulis* can often be propagated simply and conveniently with layers, pegging them down into pockets of sandy compost. *P. caerulea* and its varieties are best propagated with summer cuttings, using half-mature sideshoots 8–10cm (3–4in) long, just firming at the base and taken with a heel. These should be placed singly in small pots, using a compost of 2 parts sharp sand, 1 part loam and 1 part sphagnum peat, plus the addition of a little bonemeal. Watered well, and stood in a closed case with bottom heat of 24°C (75°F), they usually root within a month. Air should then be admitted gradually until the case is fully open, at which stage the rooted cuttings should be potted on, over-wintered in a cold frame, and planted out the following year.

Paullinia
Paullinia thalictrifolia (Sapindaceae)

A climber from Brazil with handsome evergreen foliage, ascending with the aid of tendrils to 4m (13ft) or more. The flowers, which appear in the autumn, are very small in clusters of pale pink, and its main attraction lies in the striking foliage, with its large, triangular, bronzy green leaves. The cultivated variety Argentea has silvery, rather than bronzy foliage.

As a tropical forest creeper growing by nature through and over the foliage of tall trees, but not itself climbing very high, paullinia is used to indirect sunlight. Full sunshine encourages heavy bronzing of the leaves but, carried to excess, can check the growth and cause scorching, so a balance needs to be struck between the two extremes of light and shade. Established in a container within a sunroom or conservatory, the growing tips can be pinched regularly to thicken the

Plumbago capensis has conspicuous China-blue flowers which contrast well with the handsome foliage of paullinia

foliage, stopping the leading buds on all except those shoots required to climb. Treated in this way it will make a very fine foliage houseplant, and as a permanent feature in a sunroom it will associate happily with more conspicuously flowered and comparatively easily grown climbing plants, such as the sky-blue *Plumbago capensis.*

Growing compost should be of the woodland type, based on a fibrous loam and containing a high proportion of leaf mould and peat. Watering during spring and summer should be frequent and heavy, and occasional applications of liquid feed will keep the foliage healthy. A summer growing temperature of at least 24°C (75°F) is desirable, and the winter temperature should not fall below 5°C (40°F).

For propagation, young terminal shoots about 8cm (3in) long may be taken in the spring as they start into growth, cutting immediately below a node, and these root fairly readily under thin polythene film, in a compost consisting of equal parts sharp sand and sphagnum peat, over bottom heat of 16°C (60°F).

Potato Vine
Solanum jasminoides (Solanaceae)

This beautiful plant from upland Brazil thrives outdoors in the cool summers of temperate regions, if given a sheltered spot and a light, well-drained soil; a fast-growing, slender, shrubby climber reaching some 4.5m (15ft), and flowering with great profusion from midsummer to autumn. Typically, the species produces loose clusters of pale blue flowers, but the form most commonly seen in gardens is probably the white-flowered *S. jasminoides* Album. Both types grow rapidly even in poor soil, provided they are given full sunshine at the crown.

Flowering racemes usually carry about ten individual blooms, and they also appear singly here and there amongst the bright green foliage, each star-like flower measuring some 2cm (¾in) across the petals, either a pale slate blue or white, or sometimes white flushed with mauve, but always with a prominent yellow staminal beak formed by the anthers grouping into a point – like a giant pale-flowered version of the wild British woody nightshade. Sensitive to daylight, the petals close up at night and give their best performance only in full sun, so that flowering capacity is greatly reduced when the plant is growing in partial shade. Given sunny conditions, the flowers appear in succession, often continuing until the first frost. There is a

The potato vine, *Solanum jasminoides*, gives its best performance in full sunlight and thrives on a house wall, often flowering until the first frosts of autumn

less vigorous variegated form – Variegatum – which has creamy white blotches on its leaves, and this is a fine plant, easily managed under glass.

Cool greenhouse treatment calls for little more than the exclusion of frost, maintaining an average, intermediate atmosphere, with water given sparingly during the winter. The potato vine is utterly uncritical regarding soil, asking only an open border which allows free rooting. Pruning outdoor plants when necessary should be done in the spring; under glass, the work can be done during the winter, cutting out weak, crowded growths and tying in the remainder.

Provided the site is well drained and the ground does not normally freeze more than an inch or two deep, there is no need for cosseting over winter. On a south-facing wall at the Royal Botanic Gardens in Edinburgh, *S. j.* Album thrives and blooms each summer, the yellow-centred white flowers as conspicuous as stars amongst the glossy green leaves; also in Scotland in the grounds of Tyninghame, a stately home west of Dunbar on the south-east coast, the pale blue potato vine grows on a south-facing wall and intermingles with the richer blue flowers of *Ceanothus* Burkwoodii, as the backdrop to a collection of agapanthus – the blue African lily. In Wales at Powis Castle it thrives on a corner of the terrace where it catches the morning sun to the east, while keeping company in the lee of the wall with a bright scarlet pomegranate and the autumn-flowering chaste tree, *Vitex agnus-castus.* Further south, and facing south-west under the influence of Atlantic breezes on the house walls of Glendurgan in Cornwall, England, *S. jasminoides* is semi-evergreen, bulked with the glossy, dark, evergreen foliage of the white spring-flowering *Clematis armandii* − a succession to carry the flowering season through from May to September; and at Sternfield House in East Suffolk, the white-flowered form shares a sheltered nook with a collection of clematises and honeysuckles.

Propagation may be achieved with cuttings of half-mature sideshoots 6–8cm (2½–3in) long, firming at the base and taken with a heel during the second half of summer. These should be placed singly in small pots, using a compost of 2 parts sharp sand, 1 part loam and 1 part sphagnum peat, allowing a little pure sand to drop around the base of each cutting. Bottom heat of around 21°C (70°F), very light shade, and a comparatively dry atmosphere will give the best results. As soon as growth starts they should be transferred to an open bench, potting them on as soon as they are growing strongly into a very fibrous loam. Alternatively, fair results can be obtained over winter by taking mature sideshoots 8–10cm (3–4in) long, just as growth ceases in the autumn, placing them in a sunny cold frame with a compost which might consist of 2 parts sharp sand to 1 part sphagnum peat, with a little bonemeal.

Purple Wreath

Petrea; Mexican bridal wreath
Petrea volubilis; P. racemosa and others   (Verbenaceae)

One of the most glorious of the evergreen, twining, climbing shrubs, a native of tropical America and the West Indies, with pendant 45cm (18in) racemes of deep violet-blue flowers, each individual star-like bloom up to 5cm (2in) across. In botanical terms, the true flowers are very small and dark purple, and these fall fairly soon, leaving the large lilac sepals intact on the plant; the effect of this is to make the flowers appear bi-coloured when seen from close quarters.

I once spent some months working in a tropical tree and shrub nursery − a well-organised place where large specimens of flowering shrubs were on show, and all the most striking climbers had been trained on trellises, displayed to their best sales advantage. Amongst the vivid flower colours – the many shades of crimson, scarlet, yellow, orange and blue – it was the petrea which caught the visitor's attention first. Browsing potential customers without much local gardening experience would almost invariably exclaim and enquire about the beautiful purple climber; such is its impact on the eye.

Petrea normally climbs to about 4.5m (15ft) and spreads its somewhat roughly textured foliage little more than 2m (7ft) across; it enjoys a rich soil and plenty of water and, although it will survive quite low temperatures, it needs a winter minimum of 7°C (45°F) to thrive. When subjected to slight frost during the winter, subsequent growth is liable to be slow, and flowering sparse. It manages to survive, for example, at Johannesburg, one of the coldest areas of South Africa which often experiences frost and occasional falls of snow; there, in company with yellow cassias and magenta bougainvillea, it occupies many a sheltered spot such as a north-facing suburban wall. In comparable climates within the northern hemisphere it will tolerate outdoor sites on a south-facing wall, given winter protection. A white-painted wall as a backdrop will not only display the flowers brilliantly; it will serve to reflect and magnify all the available light and warmth.

As a young plant, petrea often starts to flower while still only 60–70cm (2–2½ft) tall, especially if the growing medium is rich in decayed composted material and is kept well watered. In its native habitat the normal growth pattern is to remain moderately dormant during the cool, dry season, until the temperature rises in the spring. Typically at this season the weather remains dry, but with the increase in heat petrea starts to flower profusely, remaining in bloom until the heavy rains which normally follow within a month or two check the flowers, which reappear later to give a second display when the rains eventually subside. Most of the annual vegetative growth occurs during the wet season between flowering phases, and this pattern of behaviour suggests that overcast, humid heat and a saturated growing medium combine to reduce flowering. When grown under glass within temperate zones, vigorous growth is not as a rule required; continual flowering on the other hand is highly desirable, and it follows that a moderate but steady supply of water from spring onwards, accompanied by full light and a daytime temperature of about 27°C (80°F), not allowed to drop below 20°C (68°F) at night, will encourage the flowers to repeat without a break. This treatment also results in dark, healthy foliage and steady but modest growth. In the intermediate house at London University's Wye College, a container-grown petrea trained into the roof ties remains in flower for most of the summer months.

Other purple wreaths often confused with the typical *Petrea volubilis* include the Brazilian *P. racemosa*, more inclined to retain a shrub-like stature, scandent or sprawling rather than twining, and rarely climbing so high; this is often said to be the commonest form grown in the southern USA, in South Africa and Australia, usually classed as a distinct species, but very similar to *P. volubilis* both in flower and in leaf. It is one of the favourite shrubs in the Canary Islands where, if untrained, it remains as a rangy bush some 2.5m (8ft) high. The white-flowered variety Albiflora is usually considered to be a form of *P. volubilis*. The West Indian *P. kohautiana* has also been confused with

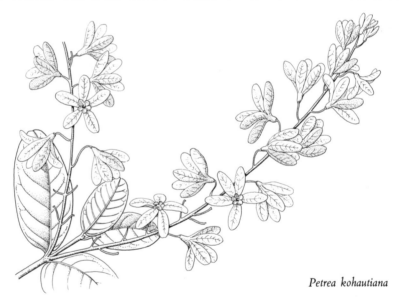

Petrea kohautiana

the other species; it is more vigorous than *P. racemosa*, with longer clusters of flowers, challenging the true *P. volubilis* in height and vigour.

Propagation may be achieved with cuttings: they root very readily outdoors in warm climates, and almost equally well under glass in temperate zones. Soft tips should be taken just after active growth starts in the spring; alternatively, the current year's trails may be used early in the summer for nodal cuttings, their length depending on the distance between nodes. These should be set close together in a very sandy compost, given bottom heat of 21°C (70°F), and a fairly open atmosphere. Roots will have formed within a month, and they should be potted up as soon as growth commences, using a compost consisting of 2 parts fibrous loam, 1 part sharp sand, and 1 part well-rotted garden compost or cow manure.

Streptosolen
Marmalade bush
Streptosolen jamesonii (Solanaceae)

This native of Colombia is an evergreen shrub normally growing about 1.5m (5ft) high, spreading to 2m (7ft), and usually taking the form of a rangy, picturesque bush. The long branches however, well adapted to scrambling over bushy vegetation in the wild, under cultivation readily adopt the scandent style of growth which enables it to be trained as a wall shrub or greenhouse climber. The dark green leaves are small, oval and roughly textured, and terminal clusters of bright orange and yellow two-toned flowers cover the bush from early spring in the tropics and sub-tropics to late winter, each bloom in the form of a narrow, twisted trumpet terminating in five broad petals, the vivid colour and glorious profusion of the flowers making the marmalade bush a personal favourite.

Though it will survive an occasional light ground frost, and tolerates a minimum air temperature of −5°C (23°F) without damage, it needs a warm, frost-free area in which to grow luxuriantly outdoors, and cannot be said to be hardy. It thrives around the coastal cities of Australia and most parts of lowland New Zealand, particularly Auckland, New Plymouth and Wellington in the North Island; and in many parts of South Africa, for instance at Pretoria, but not at nearby Johannesburg which is too high and cold. Elsewhere, even in northern temperate zones, *Streptosolen jamesonii*, overwintered under glass, can be used in exotic garden bedding as a temporary standing-out plant, well suited to any light but rich, well-watered soil, and is especially beautiful, with its lax, spreading habit, when trained as a small standard − perhaps no taller than 1m (3ft) − in which form it can be sited to dramatic effect as the centrepiece for a colourful subtropical bedding scheme. In the favoured climate of the French Riviera, the marmalade bush is one of the plants most frequently admired, and makes a perfect colour companion for blue thunbergia and yellow and scarlet tecoma.

95

Within colder temperate zones the marmalade bush is amenable to cool greenhouse treatment, whether as the occupant of a tub or large clay pot, or planted permanently in the border and trained to climb, scandent on wires or canes, or against the stays of the greenhouse roof. There it will reliably pass the winter in good health if given a minimum temperature of 4°C (40°F); heat is rarely needed during the summer months and, under these cool conditions, it will usually come into flower in the early spring and remain fully out until well after midsummer.

Streptosolen jamesonii

The marmalade bush benefits from a definite resting period, corresponding to the dry 'summer' rather than the wet 'winter' of its natural habitat, 2,000m (6,000–7,000ft) or so into the hills above the hot, jungle-clad zone of its native Colombia. Commencing at the close of summer within temperate climatic zones the compost, as well as the greenhouse atmosphere, should be allowed to dry out until it is barely moist. Flowering begins as the temperature rises, reaching its peak as the day length attains its maximum, and watering should be increased gradually in corresponding measure. High temperatures are not necessary for good flowering, but a little artificial heat is helpful to induce the buds to break if the spring weather is very cool. As soon as the flowers have opened fully further heat is rarely required, and the greenhouse ventilators should be used to limit the temperature as nearly as possible to 16°C (60°F).

For propagation, sideshoot cuttings about 5cm (2in) long may be taken with a heel, as flowering finishes during the latter part of summer. For convenience and to avoid root disturbance later, these should be set singly in small paper pots, using a compost consisting of 2 parts light loam and 2 parts sharp sand to 1 part sphagnum peat, and placed over bottom heat of 27°C (80°F). If pots of this kind are packed closely together in a fairly high-sided box, scored polythene can be fastened across the top and stretched to admit air gradually as growth commences. When the plants are about 10cm (4in) high they should have their growing tips pinched out to induce bushiness – a process which may be repeated periodically as necessary until the first flower buds form. As soon as the roots begin to pierce the paper pots they should be transferred to 13cm (5in) clay pots, using a fairly well-fertilised, sandy compost.

Thunbergia
Bengal trumpet vine
Thunbergia grandiflora
also Black-eyed Susan
Thunbergia alata   (Acanthaceae)

Thunbergia grandiflora is a magnificent perennial climber from northern India and Bangladesh, reaching a height and spread of some 5m (16ft), with 30cm (1ft) long racemes of thickly clustered tubular flowers, each 8cm (3in) both in width and length, bright mauve-blue with a pale yellow throat, the flowers continuing to open as long as the weather stays warm; in cool situations it flowers only during the peak summer months. Within temperate zones as a subject for the large, heated greenhouse, it is often in bloom continuously from spring to autumn. The large leaves are a beautiful lush green except when subjected to spells of cold weather, when they take on a blotchy grey-brown appearance.

A vigorous plant which revels in maximum sunshine and heat, in its Asian home the Bengal trumpet vine flourishes just below the mountainous regions which experience frost, and may be found in the foothills of the Himalayas, and in Nepal, where it graces the royal palace gardens in Kathmandu. It is also much admired in the Mediterranean area, and features very strongly in some of the gorgeous flower displays along the French Riviera, where its blue flowers are the perfect foil for the yellow and orange shades which often predominate. In Israel it is sometimes considered a substitute for the hybrid clematises which will not thrive there, a favourite climber for house-porch and trellis.

Black-eyed Susan is an annual or fairly short-lived perennial which climbs by twining to 4.5m (15ft). The flowers are vivid orange with a black centre, individually 2.5cm (1in) across, and numerous enough to cover the plant with colour for most of the summer. In its native South Africa it is considered a useful screen for an open fence, and a reliable ground cover, clothing the maximum area within a few weeks and flowering early. It makes rapid growth in most types of soil, and withstands long periods of drought. It will tolerate fairly severe frost too, and even when cut back to the ground will reappear in the spring and grow rapidly. In South African gardens it is usually pruned back as a matter of course, especially in mild areas where it has not been curtailed by frost; but as black-eyed Susan is short-lived and grows

Thunbergia alata, black-eyed Susan, rapidly covers a trellis. A short-lived perennial from South Africa, it is often grown as an annual and discarded when the vivid orange flowers fade

very freely from seed, it is often treated as an annual and discarded as soon as it finishes flowering.

Any moderately good soil will support the thunbergias, but the ideal is a sandy, fibrous loam to which decayed manure has been added. Given a well-drained root-run and a fairly moist greenhouse atmosphere, they grow rapidly – *T. alata* thriving in a container, *T. grandiflora* faring better when planted out in the border of a cool house and allowed to grow to its full size, removing no more than untidy or bare shoots. Although established plants can be cut back without harm during the summer, they look and flower better when permitted to cover a wide area. The leading buds are best pinched out soon after growth starts in the spring, to avoid the production of over-long shoots and encourage branching. In southern Britain and similar regions, there are many sunny nooks where thunbergias will thrive outdoors between spring and autumn, and flower well – close to a greenhouse, perhaps, against a south-facing wall that benefits from the reflected light and warmth of the glass. *T. grandiflora* needs an average temperature of 22°C (72°F) to flower well, with a controlled winter temperature of at least 16°C (60°F).

Propagating thunbergias is a simple matter in subtropical climates, using soft or half-ripe shoot cuttings taken during spring and summer; in the cool temperate zones it is trickier. *T. alata* may be grown quite readily from seed, which is usually sown in the early spring and the resultant seedlings potted up during early summer, but a succession of sowings may be carried out at any season, to produce plants ready to flower six months later. A fairly rich compost of gritty loam, leaf mould and decayed manure should be used for potting, the plants started in warmth before being planted out in a cold greenhouse. *T. grandiflora* also can conveniently be grown from seed if this is available; alternatively with cuttings taken from moderately firm trails during spring and early summer, cut below a node, with the upper leaves left intact. They should be set fairly close together in deep boxes over bottom heat of 20°C (68°F), using a compost consisting of equal parts sharp sand and sphagnum peat, and embodying a little superphosphate of lime.

Apricot
Prunus armeniaca (Rosaceae)

Difficulties encountered when growing apricots within cool temperate zones stem almost entirely from their habit of flowering very early in the year. Across much of North America as in continental Europe, seasonal change is usually well defined, but in Britain particularly, where apricot blossom opens in March or even February, spring weather as a rule is less decisive and early flowers such as these are always in grave danger from frost and cold winds. In the Mediterranean countries and in regions verging on the subtropical apricots are, of course, easy to grow; but it has often been claimed that the best and most flavoursome fruits are those produced in Britain and other countries with similar climatic features.

It is a wise precaution in such areas to make full use of the shelter offered by a sunny wall – a wall which to accommodate a well-grown apricot tree should be at least 3m (10ft) high. In south-west England the earliest varieties such as Hemskerke, and mid-season varieties like Early Orange give the best results when grown against a west-facing wall; later varieties such as Peach (a somewhat confusing name for an apricot cultivar) are best on a south-facing wall. In south-eastern England the earliest varieties are said to do best on an east-facing wall; and in the Midlands a wall facing south-west will usually give the best results. In the north of England and in Scotland apricots should preferably be grown under glass.

Paradoxically, good fruit was once easier to grow, it seems, than is often the case today. For several centuries the village of Aynho in Northamptonshire, England, supported a traditional apricot-growing industry; the fruit was marketed by the local estate which owned the village, and all the cottage walls were clothed with heavily bearing fan-trained trees, the produce of which helped the cottagers to pay for rent and upkeep. This picturesque little village is sited on a hillside facing south-west – a relatively frost-free position, with a rather poor, stony soil, described as a sandy, limy marl, and a fairly high rainfall resulting in a copious natural supply of water draining freely down the hill. While a range of apricot cultivars planted in the fertile Aynho estate gardens reportedly failed, the village trees, originally budded on wild plum suckers, were said to receive very little in the way of feeding or maintenance other than an annual pruning which took place immediately after fruiting, and normally lived highly productive lives of sixty years or more before being replaced. Their success was largely attributable to the free passage of rainwater from the cottage roofs – which in those days had no gutters or drainpipes – and through the surface of the roughly metalled or cobbled streets and footpaths. In later years, when gutters were fitted and the paths paved so that water could no longer percolate readily, the yield of apricots fell dramatically and the trees' life span was shortened to a mere twenty or so years. The variety developed from the most successful trees there was called Moorpark, and given the right conditions this is still excellent, and a reliable cultivar for outside walls. Other good cultivars include New Large Early, Hemskerke, Breda, Peach, Shipley's Blenheim, Farmingdale and Alfred, all of which ripen in the English Midlands during mid July and August.

A deep, well-drained, slightly limy soil is best. A light sandy loam is satisfactory, provided the topsoil has a depth of at least 1m (3ft); even when apricots are planted under glass, the soil in their greenhouse border should be at least 75cm (2½ft) deep. Good drainage is vital. Heavy soil intended to support apricots should be lightened with sand and bonfire ash (including charred pieces of wood), and mortar rubble is often included, giving an indication of the required pH value and free-draining texture. Bonemeal added at planting time is beneficial, and when fruiting is heavy a liquid feed can be given before midsummer. As a boost to blossom and satisfactory fruit setting, a small handful of sulphate of potash, and a smaller amount of sulphate of ammonia should be raked and watered in during late winter or early spring, before the flower buds open. A 10cm (4in) thick mulch of farmyard manure assists in ensuring even growth and, most importantly, conserves the soil moisture. Throughout the growing season the soil should be thoroughly soaked every fortnight or so; walls very often prevent rainwater reaching the roots, and dryness of the deeper soil layers is probably the chief cause of premature fruit dropping.

Well-established wall-grown trees will benefit greatly by having the topmost 2–3cm (1in) or so of their soil removed annually during the winter and replaced with new turfy loam to which has been added bonemeal at the rate of 120g/sq m (4oz/sq yd), and sulphate of potash at the rate of 30g/sq m (1oz/sq yd). Every five years or so, as a safeguard against trace-element deficiency, well-rotted farmyard manure can be used instead of topsoil.

Young trees are best planted in the early autumn, with a mulch applied immediately, not as a food but as a safeguard against drought. The roots of a young tree must never be allowed to dry out, and even at this stage apricots should receive an occasional soaking. In subsequent years some protection from cold winds and frost should be given during the crucial flowering period, commencing as soon as the buds show colour. By old-fashioned tradition, leafy evergreen branches are sometimes arranged between the branches and wall wires, to give some frost protection; alternatively, several thicknesses of plastic garden netting might be draped over the tree at this stage. In the case of apricot trees grown under glass, a close atmosphere or high temperature should be avoided early in the year, as they can readily lead to premature flower fall and crop failure. Newly planted trees in particular should be syringed frequently both before and after flowering; greenhouse red spider can become a great pest when conditions are too dry, and signs of its presence call for an increase in both syringing and watering.

The training of a fan-shaped wall tree should commence at planting time in the autumn by setting the plant about 20cm (8in) from the wall and tilted slightly towards it. At some stage, wires should be fixed horizontally to the wall about 15cm (6in) apart. During the spring following planting, the tree should be cut back to within 60cm (2ft) of the ground, immediately above a good growth bud or lateral shoot, and usually about three buds above the graft-union. From the resultant growth during the succeeding summer, two good shoots should be chosen, one on either side, each ideally about 25cm (10in) above the ground, and any buds remaining below them should be pinched off. When the two shoots are about 45cm (18in) long they should be tied in – conveniently, at this stage, to canes angled at 45° or so and fixed to the wires. Should one shoot be markedly more vigorous than the other, it can be tied down temporarily at a lower angle so that its growth will be slowed down. The main central stem can now be cut off immediately above the uppermost side branch.

During the following winter, the two lateral branches should be cut back to a bud within 45cm (18in) of their base, and the subsequent extension growths from these buds allowed to grow during the following summer and tied in progressively to the wires. Other buds will also flush along the lateral branches, and the shoots from three of these should be retained on both branches – two on the upper and one on the undersurfaces; these in turn should be trained in and tied to the wires, and all the other buds pinched off. During the next winter all eight branches thus formed should be cut back, preferably to triple buds which point upwards, so as to leave between 60–75cm (24–30in) of ripened wood on each. The following summer the end buds will give rise to extension growths which should be tied in to the wires, and the remaining buds should be allowed to form shoots, ideally spaced every 15cm (6in) or so along the branches on both upper and lower sides, tying them in place at the end of summer so that eventually they lie at an angled 10cm (4in) apart, in no case allowing them to exceed 45cm (18in) in length. Unwanted buds should be rubbed out.

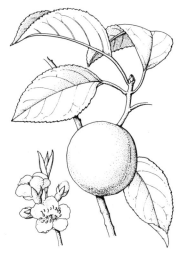

Prunus armeniaca

Apricot flowers, which are self fertile, appear on spurs of both young and old wood, and the formation of these fruiting spurs can be encouraged once the young tree has been trained to shape. Sublateral shoots should be stopped in the spring as they reach a length of 8cm (3in), by pinching out their growing tips – if all-round growth is particularly vigorous that season they can be allowed to make 15cm (6in) of growth before stopping; about one month later all the sideshoots growing from them should, in their turn, be stopped having made one leaf each, and these will form the fruiting spurs for the following year.

In Britain, apricots flower as a rule too early for bees to pollinate them, and this needs to be done by hand: a soft brush or, by tradition, a rabbit's tail is brushed lightly over the flowers in turn, ideally at midday. Should the weather be very dry, a light syringing after pollination is beneficial. When the developing fruit has reached a diameter of about 1–1.5cm (½in), a first thinning should be given, removing all those which are obviously overcrowded and touching. This operation should be repeated progressively until the ripening fruits are about 12cm (5in) apart, but the final thinning should be postponed until the stoning process is complete. Stoning is the period after the initial swelling during which the apricot stone is developing; growth is inward, and expansion appears to be at a standstill. Should the process of stoning fail, the fruit will drop off without further change; if it has been successful, a second swelling takes place, and the fruits gradually assume their full size and shape. Normally the weather is still reasonably cool while stoning is taking place and, under glass, it is safer if the temperature can be kept below 7°C (45°F) until the process is complete. Once the final swelling has commenced it can be allowed to reach 10–13°C (50–55°F), but this must not be exceeded even then or the fruit may fall prematurely. Apricots which have been allowed to ripen fully on the tree have a better flavour than those picked early and stored. As ripening approaches in the second half of summer, leaves which are shading the fruit should be removed to admit maximum sunlight. No pruning other than this should normally be carried out after midsummer, once the young tree has been trained, or the new growth which results will have no time to harden thoroughly before winter sets in.

The main work of pruning should be completed during the winter, after leaf fall and before the flower buds start to expand, and all the past season's new growth should be tied in to the wall wires. An exception to this rule occurs when leading shoots have been retained to form extra extensions, and the curtailing of these can be done during late summer. Apricots grown as free-standing bushes call for a slightly different pruning regime, entailing cutting back in the late spring all the new shoots to within a leaf or two of the swelling fruit. Bush apricots without the protection of a wall are unlikely to need watering except during periods of drought. To form an orchard they should be planted about 4.5m (15ft) apart, and as there is really no practical way of protecting the flowers in early spring, the site should be as sheltered and free from frost as possible. Apricot wood is very brittle, and high winds can also cause much damage.

Apricot cultivars are normally propagated by budding in midsummer, using seedling apricot or plum stocks. Buds to avoid as scions include those taken from particularly vigorous shoots; double or triple buds – which embody one pointed growth bud and one or two rounded blossom buds – give the best results, and to avoid desiccation the operation should preferably be carried out whilst the weather is overcast. Stock plants should ideally have a diameter of 8–10mm (⅜in) at the point just above ground level where budding is to take place, and three-year-old plants are usually ideal. Shield budding is the most convenient method, using a standard T incision, and the stock plant can be headed back as soon as scion growth is well advanced. As a general improvement measure for old trees that are no longer particularly productive surplus blossom buds from other apricot trees can easily be grafted on to the branches in the same way where-ever they are needed.

Aubergine
Eggplant; bringalls; brinjal; mad apple; melongene
Solanum melongena esculentum   (Solanaceae)

This is one of the most widely grown vegetable-fruit plants throughout the tropical world, and very popular in southern Europe, but often neglected by gardeners in colder countries in the mistaken belief that it needs tropical conditions to grow successfully. Eggplant is a herbaceous perennial growing about 1m (3ft) high, with lobed, hairy leaves and 4cm (1½in) yellow-centred violet-blue flowers. The smooth, shiny fruit varies greatly both in size, averaging about 15cm (6in) long, and in colour, with many different varieties under cultivation in purple, black, yellow or white. Some of the best cultivars selected for their good results when grown in temperate countries include the large, almost black-fruited Baren, bred for its resistance to the virus diseases which the common eggplant shares with tobacco and cucumber; also the rather more tender Marfa, which has enormous dark purple fruits up to 25cm (9in) long. These cultivars are both first-generation hybrids which cannot be expected to come true from any seed they may produce; the original controlled cross

Solanum melongena esculentum

pollination has to be repeated for each batch of seed which, for this reason, tends to be fairly expensive.

When grown permanently outdoors without protection in frost-free regions, aubergines need a well-drained sandy loam with a preferred range of acidity between pH 5 and 7, and are normally limited, even within the tropics, to an altitude below 900m (3,000ft). Elsewhere they can be surprisingly ornamental when planted and trained to cover a sunny wall or trellis, but within cool temperate zones, like the tomato, the fruit will ripen well only during a hot summer. In such climates they need a clear six months of the year wholly free from frost, and of these at least three should have a minimum temperature above 10°C (50°F). The climatic ideal would be a showery growing season with plenty of sunshine, and cool nights during which the temperature drops to between 15 and 20°C (60 and 70°F). In the tropics, excessive rain and over-warm nights often result in luxuriant foliage but poor fruiting.

Provided the seedlings can at least be started in warmth, even an unheated greenhouse will be adequate in cool temperate regions to provide a shield against the more extreme variations of temperature, to safeguard against frost, cold winds and the risk of waterlogging, and to magnify the early spring sunshine. When aubergines are to be grown in the open within temperate zones, planting out should be delayed until a month after the final frost is safely past. In an average season, a moderately heated greenhouse will allow growth to start up to a fortnight before the last frost is due.

Seeds may be sown in any commercial seed-sowing compost –
ideally it might consist of 2 parts medium loam to 1 part sphagnum
peat and 1 part sharp sand, with the addition of a little
superphosphate, but the wholly peat-based composts are suitable,
covering the seeds 1cm (¼–½in) deep, and setting the seedbox over
bottom heat of 18°C (65°F). The plants grow better for an early start,
and sowing may be carried out during the winter some two months
before the last frost can be expected. The seedbox should be covered
with a sheet of paper over the glass, but as soon as germination takes
place the seedlings must be given full light. They may be potted as
soon as they are large enough to handle easily; 8cm (3in) paper pots are
ideal for aubergine plants at this stage, and the compost might consist
of 2 parts loam, 2 parts sphagnum peat and 1 part sharp sand, to each
large barrowload of which has been added a double handful of
bonemeal, a small handful of superphosphate of lime, and a very small
handful of potassium sulphate. After potting on, the plants may be
grown to the fruiting stage in 15cm (7in) pots, but as a rule they
perform better when planted direct into a greenhouse border, with a
lumpy compost of loamy soil containing garden compost and leaf
mould, fibrous peat and sharp sand. For convenience, satisfactory
aubergines can be grown in the polythene bags of fertilised peat
intended for tomato culture.

The soil inside a greenhouse is best brought to a temperature of
about 15°C (60°F) before planting, either by heating the house for a
week or two previously, or by watering the beds with water warmed
to 27°C (80°F). Drainage in the greenhouse beds must be perfect to
enable copious supplies of water to be given while the fruit is swelling,
without the compost becoming saturated. When the plants are 15cm
(6in) high the growing tips should be picked off to encourage them to
bush out and produce the sideshoots which will bear fruit. Eggplants
should be spaced about 75cm (30in) apart, and top-dressings of strawy
manure or garden compost are very beneficial. Fruits should be limited
to about six per plant if they are to attain a fair size. The compost
should be no more than moist until the fruits start to swell, at which
stage watering is increased and the plants fed regularly with a mixture
consisting of 2 parts sulphate of ammonia, 2 parts sulphate of potash,
and 3 parts superphosphate of lime, allowing one heaped teaspoonful
of this dry mixture per plant weekly in a heated greenhouse, or every
ten days in an unheated one, sprinkling it in a circle around the roots
and watering well in. Commercial high-potassium tomato fertiliser

will do equally well at this stage, with a choice of dry or liquid types.

Although eggplants are perennial, they are normally grown as annuals. Cuttings can be taken in the autumn and the resultant plants overwintered; alternatively, they may be taken under glass around the time of the final spring frost in cold temperate climates, and they will then bear fruit later the same year. Seedlings are usually best and most reliable. Should it become necessary to produce additional plants from cuttings, these can be secured as soon as the sideshoots appear, removing them below their basal bud with a wedge-shaped cut and rooting them in an equal parts peat/sand compost, over bottom heat of 21°C (70°F).

Banana
Musa paradisiaca sapientum; M. cavendishii and others (Musaceae)

Most of the bananas of commerce are varieties of *Musa paradisiaca sapientum*, very widely grown in many hot countries throughout the world, and needing tropical conditions in which to thrive. Far hardier, but just as good to eat, is the so-called Chinese banana, *M. cavendishii*, an attractive, thin-skinned, seedless and delicately flavoured fruit. This banana species is quite widely planted outdoors in the subtropics and the warmest temperate zones. Growing no taller than 1.5m (5ft), with handsome greyish-green leaves up to 1m (3ft) long and 30cm (1ft) wide, the Chinese banana is a useful plant in temperate countries for the cool greenhouse, if there is enough room to house it, comparatively easy to grow in a large container, where it can make a spectacularly ornamental subject which will produce a bonus of fruit from time to time. Under these circumstances it needs a minimum winter temperature of 7–10°C (45–50°F), and a minimum growing temperature of 16°C (60°F).

Whether planted in a tub or rooted permanently in the greenhouse border, *M. cavendishii* needs feeding fairly richly during the growing months, and appreciates a fertile loamy soil to which farmyard manure has been added. Very little water should be given during the winter but, like all bananas, it benefits from copious supplies during the summer months, coupled with a very humid atmosphere – brought about by arranging containers of water in the greenhouse, by regularly damping down the surrounds, and by syringing the plants themselves. The water used for syringing and watering should for preference be lukewarm, and will be most effective if applied only during the after-

(*Left*) *Musa paradisiaca sapientum*, the banana of commerce, needs tropical conditions in which to thrive. (*Right*) The hardy Chinese banana, *Musa cavendishii*, can make a spectacular ornamental subject which will produce excellent fruit from time to time

noons with the greenhouse ventilators closed.

As the fruits are seedless, for propagation it is necessary to rely on the suckers which arise prolifically from stolons below the base of the stem. These suckering shoots may be removed at any time as they reach a convenient size, potted individually into a very rich soil, and encouraged to grow as vigorously as possible by maintaining a high temperature and humidity. The main flowering stem dies after fruiting, but suckers normally appear before this occurs. The time taken for a new sucker to come into bearing varies considerably, but averages about eighteen months after planting. Fruiting stems of the Chinese banana are likely to be at least 1.25m (4ft) and sometimes 1.5m (5ft) high, and though the greenhouse can be fairly small to accommodate them, it will need a roof ridge at least 1.75m (6ft) in height, or the glass is liable to be broken as the leaves expand.

In the Canary Islands, the Chinese banana is very popular and grows everywhere in gardens and orchards with other fruit such as mango, pineapple, citrus and avocado; and in Florida and other south-eastern

United States it is grown as much for its decorative, architectural value as for its fruit. The more tender *M. paradisiaca* varieties are to be found, as a rule, only under glass in North America – as in the remarkable tropical fruit house at the Longwood Gardens in Pennsylvania, where tall banana trees grow in the company of papayas, festooned with fruiting granadillas, above tea and coffee bushes at their feet.

Among other banana species, red-leaved varieties of *M. ensete*, the Ethiopian banana, are used as summer bedding plants in places with long, warm summers – California for instance, and the south-eastern USA, New Zealand and the south of France, where they are usually grown permanently in tubs containing a richly manured compost and plunged outdoors in the spring. One of the hardiest of ornamental bananas is *M. basjoo*, which produces only tiny fruit, and is sometimes grown as a subtropical bedder within temperate countries, at places such as seaside resorts in the milder English counties, where these plants can make an impressive thicket of arching 3m (10ft) leaves, each some 60cm (2ft) wide, of a bright shiny green, but they need a site sheltered from sea breezes, as the leaves are easily torn to shreds in high winds. *M. basjoo* is hardy enough to remain outdoors for the winter in south-west England, provided the crowns are given some protection in the form of a portable frame, or a polythene sheet enclosing bracken or bonfire ash.

Calomondin
Citrus mitis (Rutaceae)

This attractive little shrub, a native of the Philippine Islands, has become popular as a house or conservatory plant for its decorative value alone. The white, fleshy-petalled sweet-scented flowers are out on the thornless twigs often for eight months of the year, and when grown as a pot plant, a large specimen will usually have a few flowers or fruits on display at most times of the year, often with green, ripening and fully ripe miniature oranges to be seen simultaneously. Quite apart from its undoubted ornamental value, it can be grown with intensive cropping in mind, for these dwarf citrus fruits are excellent for eating raw, for crystallising, fermenting, or making into soft drinks. Green at first, ripening to orange-yellow, the 3cm (1¼in) fruits have an aromatic but slightly acidic flavour when ripe, are very thin skinned, and contain usually only a few tiny seeds. Given the correct treatment well-grown calomondin plants no more than 1m

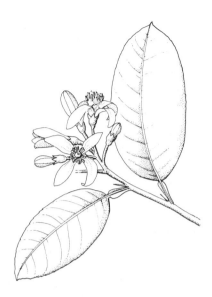

Citrus mitis

(3ft) tall can carry as many as two hundred fruits at a time.

Within temperate zones calomondin needs to be given a minimum temperature of 10°C (50°F), and a position in the sun, near the glass for maximum light in the winter and, if possible, outdoors during the summer. Pot-grown plants managed for fruit production benefit from an organic feed, such as a dilute solution of bloodmeal, given little and often, and should be sprayed annually with a colloidal copper fungicide, and periodically throughout the year with a foliar feed containing trace elements. Clay pots should be used in preference to plastic – chiefly for their greater stability when the well-developed plants become top-heavy – and both pots and compost should be very free draining. Dryness at the roots quickly results in shrivelling and dropping of the leaves, and watering should be thorough. A peaty potting compost that has been fairly well fertilised is suitable, for example 3 parts loam and 4 parts sphagnum peat to 1 part sharp sand, with the addition of a large handful per barrowload of a fertiliser made up of 2 parts superphosphate of lime, 3 parts hoof and horn meal, and 1 part sulphate of potash.

The fruits which ripen during the late winter are those most likely to contain plump, fertile seed. Seed intended for sowing should be strained from the pulp and soaked in fresh water for two days, then sown at once, using a compost consisting of 2 parts loam, 1 part sharp sand and 1 part sphagnum peat, to which has been added a little superphosphate. Given bottom heat of 21°C (70°F), it usually germinates quickly and well. Calomondin seedling plants grow more

111

The calomondin, *Citrus mitis*, can produce heavy crops of miniature oranges, and makes a good houseplant

vigorously than rooted cuttings, but they are spinier and seldom flower for the first six months; rooted cuttings produce a fair quantity of fruit, but grow rather poorly; grafted plants do better than either – they grow strongly and vigorously, and produce heavy yields of fruit, often during every month of the year, especially in the autumn and early winter. When intended purely as an ornamental house plant, therefore, calomondin is usually grown on its own roots and the plants are shapely and attractive, but the root system is normally too weak for heavy bearing. Grafted plants grown for fruit production, on the other hand, are strongly upright in habit, and it is advisable to encourage branching by pinching out the growing points regularly as they develop. The root growth being more vigorous, such plants are liable to become pot bound unless potted on as necessary each autumn.

Scions selected from fertile, non-spiny clones should be grafted on seedlings to give the best fruiting results. The seedlings should be established in 13cm (5in) pots, and are at their optimum development for grafting at about eighteen months old, when they should be around 38–45cm (15–18in) tall. Early summer is the ideal season to graft; the resulting union will then be complete within a month, and the new plants should commence bearing towards the end of the following year. Stock calomondin seedlings should be beheaded 7–10cm (3–4in) above soil level, and a vertical slit made about 2.5cm (1in) deep in the cut surface, ready to receive the scion, which is trimmed at its base into a narrow wedge shape. The scion ideally should be about 8cm (3in) long, and must be inserted so as to match the cambium exactly, preferably on both but at least on one side of the graft. Soft polythene tape can be used to tie, and the grafted plant should be enclosed in a polythene bag held apart with strategically placed twigs, and secured by an elastic band around the pot, which is then placed in a shady spot under glass. New growth should have appeared after a month, at which stage the polythene bag can be partly opened either by cutting or raising the base, removing it completely after a further week. The polythene tie around the union should be left untouched until stock and scion are obviously united firmly and general expansion is taking place.

Citrus
The larger *Citrus* species   (Rutaceae)

Cultivated since ancient times, the large citrus trees are probably natives of the south-east Asian monsoon regions, where the seasons alternate beween wet and dry. They do not grow very well in humid areas such as the tropical rain forest regions of the world. Most of the very numerous varieties and hybrids now in cultivation are to be found in those subtropical zones which have the Mediterranean type of climate, sometimes grown with the aid of seasonal irrigation. Their outdoor range lying mainly between the latitudes 45°N and 35°S, the bulk of the commerical crop is produced between sea level and 600m (2,000ft), and even near the equator citrus does not do well over 1,800m (6,000ft). Day length seems to have no bearing on their performance.

Provided they are truly dormant at the time, most of the cultivars will tolerate light frost for short periods; but cold spells, even above freezing, can be harmful during the growing season. Citron and lime are the most frost-tender of the group; others, progressively more hardy, are lemon, grapefruit, sweet orange, sour orange, mandarin, kumquat, citrange and, hardiest of all, the trifoliate orange. In the case of the more tender kinds, vegetative growth is greatly reduced below 13°C (55°F), but they will tolerate high temperatures reaching 38°C

(100°F) or more. If grown without irrigation, all need an annual rainfall of at least 88cm (35in). A 1.5m (60in) rainfall is rather too heavy, though mandarins will tolerate wetter conditions than the others.

As a commercial crop, citrus will thrive in a wide range of soils, the ideal averaging a light, fertile loam with a pH ranging a degree or two on either side of neutral, but various clones have been selected for use in more extreme conditions. Typically they develop a strong taproot and a mass of surface feeding roots, and the shoots and roots tend to grow in cycles: while top growth is dormant the roots are active, and vice versa. The sweetly scented flowers usually appear in the subtropical spring, and fruits take from seven to fourteen months to ripen after pollination. Only a small proportion of the flowers actually produce fruit, and many which start developing drop off during the embryo stage. Pollination is effected by bees and other insects, and most varieties are both self- and cross-compatible.

Citrus leaves which show signs of yellowing or bronzing are frequently displaying symptoms of magnesium deficiency, and will benefit from an application of Epsom salts. Deficiency problems can be avoided by spraying occasionally with a foliar feed containing trace elements, and a mulch of farmyard or stable manure is also advantageous. Citrus trees when grown in a small way in the subtropics are often intercropped with vegetables so as to benefit from the routine fertilising.

The sweet orange *Citrus sinensis* is the most widely grown species, with numerous named cultivars. The most important of these in the world's markets include Jaffa, which grows well in Palestine; Valencia, a spring-ripening variety which does well in the West Indies and the Mediterranean area; and the winter-ripening Washington Navel which, with Valencia, is important in California, parts of South Africa, and Australia.

Satsumas, mandarins and tangerines are all varieties of *C. reticulata* (syn. *C. nobilis deliciosa*): satsumas, extensively grown in Japan, are very hardy and cold-resistant; mandarins – pale orange to yellow in colour – include the popular cultivar Emperor which is widely grown in Australia; tangerines – deep orange-red – include Dancy, very popular in the United States.

A hybrid derived partly from the sweet orange has been named the citrange – fairly hardy outdoors in cooler parts of the USA and the warmer districts of Britain. There are several varieties of citrange,

Citrus sinensis. The many varieties of sweet orange usually take from seven to fourteen months to ripen after the fragrant flowers appear

ranging from reddish berries to pale orange fruit 10cm (4in) across, interesting for various household purposes, though never as sweet as the true sweet oranges. As a flowering shrub citrange is quite attractive in the spring, with fragrant white blooms measuring 6cm (2½in) across the narrow, star-like petals.

Hardiness in the case of grafted plants depends in some degree on the rootstock used for grafting or budding. Sweet oranges grafted on to sour orange stocks are hardier than if grown on their own roots, and the use of a rootstock which is faster growing and more vigorous than the scion usually results in an earlier fruiting, dwarfer tree with well-developed roots. The sour orange, *C. aurantium*, makes a useful stock in the Mediterranean countries and in Florida, where its deep taproot makes it resistant to drought, and equally able to flourish on heavy, damp soils. As a budded rootstock, the less hardy sweet orange is subject to frost damage and grows best in rich, well-drained loam, making a good stock in California. The wild or rough lemon, which

Poncirus trifoliata

has become naturalised in many parts of the world – it is often to be found growing wild in the bush of Central and East Africa – is less hardy than either of the oranges, being greatly subject to frost damage. It grows best on a light, sandy, well-drained soil, and makes a good rootstock in South Africa and Australia.

The hardiest rootstock for use in the colder districts is the trifoliate orange, also known as the golden apple or Japanese bitter orange, *Poncirus trifoliata.* Trees grafted, or budded, on this stock are moderately dwarf, early-bearing and prolific, and tolerant to some degree of frosting and waterlogging. As a budding stock it is compatible with satsumas, mandarins, tangerines, oranges and grapefruits. *P. trifoliata* is an attractive garden shrub in its own right, hardy in Britain and the eastern USA, where it needs a well-drained soil and a site in the sun. It forms a slow-growing bush eventually reaching some 4.5m (15ft), with green stems formidably armed with long spines, and sweetly scented white orange blossom in the spring, each flower some 5cm (2in) across the petals, followed by fragrant, downy, 5cm (2in) yellow fruits. In southern Europe it is often used as a hedging plant. The seeds ripen, even in England, and may be used for propagation. Alternatively, cuttings of half-matured shoots will root if taken soon after midsummer and kept close in a sandy compost over bottom heat of about 27°C (80°F).

Cultivars rarely, of course, come true from seed, and named citrus varieties are normally budded on seedling stocks. Seeds should be washed free of all traces of pulp and juice, and sown before they are completely dry – in the subtropics, outdoors in well-drained, shaded beds, elsewhere, under glass in seedboxes or pots containing a compost

116

consisting of 2 parts light loam, 1 part sphagnum peat and 1 part coarse sand, embodying a little superphosphate and a sprinkling of ground limestone. Citrus seeds will not germinate below 13°C (55°F), and the optimum temperature range for even germination is between 27°C (80°F) and 32°C (90°F). Seedlings are best potted up when about 20cm (8in) high (which in the subtropics will be about six months after sowing), and budding is normally carried out about six months later.

Citrus shoots are characteristically angular when young, and it is not until they become rounded in section that they are ready for use as budding scions. The buds should be taken while still dormant in the very early spring, and the stock plants need to be of about pencil thickness 30cm (1ft) from the ground. As it is desirable for the stock to be slightly more advanced than the scion, budwood which is taken while still dormant can be stored in damp sawdust, keeping the shoots in good condition until the stocks are in active growth. Traditionally, an inverted T-shaped incision is used, and the bud bound into place with raffia or, more conveniently, transparent plastic tape. If the bud is still green two or three weeks after budding, the take will have been successful and the tape can be removed. If the bud is found to be dead at this stage, another attempt can be made. When the new scion shoot has reached a length of 2–3cm (1in), the stock should be cut off immediately above the union. As the scion grows it should be staked and progressively tied, and eventually topped at about 1m (3ft) to induce branching, any suckers being removed.

Poncirus trifoliata, known as the Japanese bitter orange or golden apple, is hardy and makes an attractive slow growing shrub or thorny hedge

Ficus carica. In frosty temperate zones the figs which develop near the tips of the shoots should be retained over winter, and will ripen the following summer

Ficus carica. A wall-trained fig tree rejuvenated by cutting back the old wood and tying in the resultant young shoots. It will take a few years to come back into bearing

Fig
Ficus carica (Moraceae)

Originally a wild plant of the Mediterranean region, since antiquity the fig has been very widely cultivated in warm countries everywhere. In the USA it is grown extensively as a commercial crop in California, Texas and Louisiana – and with winter protection as far north as Pennsylvania. In southern England it will grow as a standard fruit tree, but in the north of the country and in Scotland it needs the protection of a sheltering wall.

Although deciduous by nature, the fig bears continuously – two or three crops a year if the climate is suitable – and difficulties of cultivation arise only in areas where figs are on the verge of tenderness. In temperate climates it is enough if only one crop is produced. As growth progresses in the spring, embryo fruits appear in the leaf axils and, later in the season, towards the tips of the new shoots. In most cool temperate climates, the new year's growth cannot begin until after the final frost, and the growing season is not long enough for the current year's figs to develop and ripen before the first frosts of the autumn spoil them. To be successful in regions such as this they have to be grown on a biennial basis, and those fruits which appeared later in the season need to survive the winter if they are to ripen properly the following year. To allow the ripening of three crops a year, the temperature in midwinter must remain in the region of 18–21°C (65–70°F), gradually rising to 27°C (80°F) as spring approaches; for two crops to develop satisfactorily during the year, the winter temperature needs to average around 12–15°C (54–59°F), again rising to 27°C (80°F) early in the summer.

On trees growing in the open within frosty temperate zones, figs which set early in the year and have reached a fairly well-developed stage by the autumn stand little chance of ripening but, instead, will drop during the winter – they are best picked off during the spring, thus saving the tree's energy. Those embryo fruits which appear later during the summer, however, usually near the tips of the current year's shoots, should be retained and will remain safely dormant through the winter, ready to start growing in the spring. Their formation can be encouraged each summer by pinching out the shoot tips at their fourth or fifth bud, at the same time removing the early fruits already forming towards their base. Ideally, the bearing shoots will be short jointed and about 20–25cm (8–10in) long; sappy shoots which start into growth late will not harden sufficiently to withstand the winter, and should be removed completely. When only one crop a year is to be grown, ripening of the fruit commences during the second half of summer and continues well into autumn if the weather remains mild. For prime quality, figs should be allowed to ripen on the tree; ripeness is indicated when the stalks droop, allowing the fruit to hang down limply.

Good cultivars suitable for growing outdoors in temperate climates and which can also be forced into early ripening under glass, include White Marseilles and Brown Turkey. The former produces large, almost round, thin-skinned, pale yellowish-green fruits, nearly white when ripe and with almost transparent flesh, rich and sweetly flavoured; the latter is a more typical fig, fairly large and pear-shaped, brownish purple when ripe, the inner flesh tinted reddish, with a rich, sweet flavour.

Soil in which figs are growing should not be too rich. A light sandy loam containing old mortar and brick rubble is ideal, with a little lime and wood ash added; heavy soil encourages too-lush foliage at the expense of the fruit. Copious water must be given from the time growth starts in the spring, and if the soil is allowed to become dry the developing fruits are liable to drop off before they can ripen. To encourage heavy fruiting the roots of a wall-trained plant should be restricted, either by regular root pruning with a spade, or by boxing them in with a concrete trough about 75cm (2½ft) deep by 90cm (3ft) wide, a bed 1.5m (5ft) in length thus holding about 1cu m (35cu ft) of soil. Ample drainage holes should be provided and provision made to ensure they do not become blocked. Young trees should be planted in the spring and well firmed in, mulching immediately with well-rotted manure to help retain moisture in the soil.

As new shoots appear from the base, they should be tied in to the wall to form a fan shape, using wires spaced about 22cm (9in) apart. The wall should be at least 2.5m (8ft) high to provide adequate support and protection for a fig tree, which will take three or four years to form the full-sized fan. Once a fig tree is well established, although on its own roots, sucker growths are best removed as they also take a few years to come into bearing. In districts where the weather is likely to be severe, wall plants should be protected during the winter by covering or thatching with straw or bracken – traditionally often done by untying all the shoots at leaf-fall and bundling them gently in straw. Pruning can conveniently be carried out in the early spring, removing dead, damaged and surplus shoots and overcrowded branches.

Propagation is normally achieved with cuttings of ripe wood, usually about 10–12cm (4–5in) long, taken with a heel during the winter. They should be placed singly in small clay or paper pots in an open frame, over bottom heat of about 24°C (75°F), using a compost consisting of 2 parts fibrous loam, 1 part sphagnum peat and 1 part coarse sand. If paper pots are used they should be crowded closely together so as to eliminate air spaces, but the old-fashioned clay type are best plunged up to their necks in sand.

Golden Berry
Cape gooseberry; Peruvian cherry
Physalis peruviana edulis   (Solanaceae)

A somewhat floppy bush 1–1.5m (4–5ft) high with a spread of 2–2.5m (7–8ft), the very hairy stems of this Peruvian herbaceous perennial bear slenderly long pointed heart-shaped leaves, with white violet-anthered flowers appearing in the axils during midsummer, followed by delicious yellow gooseberry-like fruits which develop inside dried calyces like greenish-brown Chinese lanterns. Allied to the potato and the tomato, it is a close relative of the much smaller, hardy ornamental garden plant *Physalis alkekengi*, the bladder cherry or Chinese lantern, whose bright orange-red calyces are often used by flower arrangers.

Some individual golden berry plants habitually bear far heavier crops than others, and work is being done on breeding and selecting the best clones. Meanwhile, if given a choice of propagating material, only those plants known to bear prolifically should be used; the small-berried kinds as a rule produce the heavier crop, and their fruit is said to have a better flavour.

When grown outdoors within temperate regions, even a mild frost will cut back the shoot tips and a severe frost will kill the plants, but good crops can still be had outdoors. Established in containers, golden berries fruit very well during the summer and can be carried inside before the first frost is due, to continue fruiting under glass throughout the winter, given a minimum temperature no higher than 10°C (50°F).

In the wild, golden berries inhabit comparatively poor hill soils, and their growing medium under cultivation should not be too rich;

Physalis peruviana edulis. Golden berries may reach 3cm (over an inch) in diameter, but the smaller fruits tend to have the best flavour

planting in open garden beds tends to encourage over-vigorous vegetative growth, and fruiting under these circumstances is often poor. Established permanently in 20cm (8in) pots they give excellent results if plunged in the soil, either outdoors or in the greenhouse border according to season; when grown in containers, however, they are best cut back to 1m (3ft) or so when brought under glass for the winter. Plants grown as permanent residents in the greenhouse lend themselves readily to training against a wall or wire framework, and outdoor plants too are best supported with canes or wires to prevent them flopping over. When tied upright they can reach a height of 2m (7ft), and grown thus they may be spaced at 75cm (2½ft) intervals along rows conveniently 1m (3½ft) apart. In spring, overwintered container-grown plants are best persuaded to start into growth under glass before being carried outside after all risk of frost has passed. Alternatively outdoor frames or portable cloches can be used to start closely trimmed plants into growth, an early start such as this resulting in a greatly increased yield. Should shortage of storage space under glass not allow overwintering of large plants, small rooted cuttings taken in the autumn can easily be kept free from frost, and started in warmth early the following year ready to set outside as soon as the weather allows.

Golden berries are ready to pick when the calyx has turned from green to amber and feels dry and papery to the touch, and they should be picked with the calyx intact. Provided they show some degree of yellowing, the berries will continue to ripen if kept in a light, warm place – eg spread thinly on a tray in a living-room window.

Propagation may be achieved with seed or cuttings. Seeds (saved when the berries are being prepared) should be sown very early in the spring, using a compost of 2 parts sphagnum peat to 1 part loam and 1 part sharp sand, covering them 5mm (¼in) deep and placing a sheet of glass over the container, using bottom heat of 18°C (65°F). As soon as they are large enough to handle, the seedlings should be pricked out 3–4cm (1½in) apart into a seedbox, and transferred to individual paper pots or polythene bags when about 5cm (2in) high, using a compost consisting of 3 parts loam, 2 parts sphagnum peat and 1 part sharp sand, incorporating a general fertiliser at the rate of one medium handful to the large barrowload. A suitable fertiliser might consist of 3 parts hoof and horn meal, 2 parts superphosphate of lime, 1 part sulphate of potash and 1 part ground limestone. If flowers have not appeared by the time the transplants are 30cm (1ft) high, the tips should be pinched out to promote bushiness, and they can be planted outside any time after new growth appears. Alternatively, the seeds may be sown after midsummer, using the same method but overwintering the resultant plants in a frost-proof cold frame before planting them out the following spring.

Plants grown from cuttings crop earlier and are less vigorous than seedlings, so tend to give the best results when planted direct into open ground. Cuttings will root fairly readily at any time of the year, and it is often convenient to take them in the autumn as the top growth is being cut back and the plants are housed. Shoot tips 10–15cm (4–6in) long should be used, cut below a node, and set in a compost consisting of 2 parts loam and 2 parts sharp sand to 1 part sphagnum peat, incorporating a little superphosphate and ground limestone, setting over bottom heat of 21°C (70°F) but keeping the tops open and well aired. They should have rooted within a fortnight and be ready to pot after three weeks, using the same potting compost as that described for seedling transplants.

Granadilla
Passion fruit; passion vine
Passiflora edulis and others   (Passifloraceae)

This native of Brazil is widely grown for its fruit throughout the tropical and subtropical regions of the world, particularly in America,

Africa, Hawaii, Australia and New Zealand. Many of the passion flower species bear edible fruit, but the 4–5cm (2in) egg-shaped granadilla of the typical *Passiflora edulis* bears more prolifically and has a better flavour than most. The typical dark purple form of the species is also said to have a pleasanter and rather less acid taste than the other naturally occurring and equally widespread variety, the yellow-fruiting *P. edulis flavicarpa*; but both kinds are popular wherever they grow well, and their local distribution depends to some extent on micro-climate and topography: both prefer a rainfall in the region of 80–130cm (30–50in) per annum; but the typical purple form does best above the 1,800m (6,000ft) contour in the tropics, and is less well suited to the wetter, hotter lowlands; the yellow form on the other hand seems to prefer the tropical lowlands where it thrives on the heat and humidity.

In North America the wild species include the vigorous passion vine, *P. incarnata*, also known as May pops or May apple, with lavender-blue flowers and 5cm (2in) yellow fruit; and the smaller and less vigorous yellow passion flower, *P. lutea*, which has purple fruit – both species are to be found in the wild from Virginia to Missouri, and south to Florida and Oklahoma. A few other wild species with edible fruit also occur in the southern and western United States.

Other passifloras sometimes cultivated for their fruits include *P. laurifolia*, the water lemon or pomade de liane, also known as the Jamaican honeysuckle, which grows wild in thickets and on forest

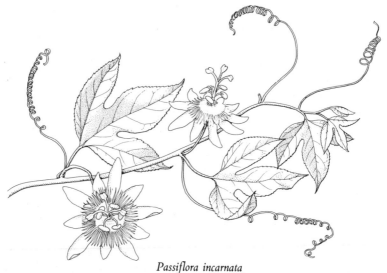

Passiflora incarnata

124

fringes in the West Indies and adjacent parts of South America; valued for its handsome, fragrant flowers in pink and violet, as well as the egg-shaped orange-yellow fruit, it has become widely distributed throughout the tropics of the world since first being carried abroad in the seventeenth century. It grows well in humid lowland regions, and has proved the most suitable passion fruit for growing in Malaysia. *P. ligularis* is the sweet granadilla from tropical America, with tawny-orange egg-shaped fruit, the best species for growing in the mountains of Mexico, Central America and Hawaii. *P. molissima* is the bananadilla from the Andes, with pink flowers and downy yellow 7cm (3in) banana-shaped fruit. This is best suited to cooler conditions, thriving in New Zealand and the colder parts of South Africa, and has become naturalised in several mountainous countries, particularly in Hawaii, where it abounds around the 1,500m (5,000ft) contour on the volcanic slopes. *P. antioquiensis* from Colombia is similar to *P. molissima*, but with bright red flowers; it also does well at high altitudes, and is hardy enough to grow out of doors in the Scilly Isles. *P. quadrangularis*, the giant granadilla from tropical America, does best in hot, moist lowland regions within the tropics, enjoying the same conditions as the yellow-fruited variety of *P. edulis*. The cylindrical fruits of the giant granadilla measure up to 25cm (10in) long, pale yellow when ripe, juicy but not particularly flavoursome. When grown under cultivation the flowers usually need hand pollination for fruit production, but it is rarely used as a commercial crop in the tropics. A vigorous perennial climber, under glass it may be grown on overhead trellis work, allowing the fruit to hang beneath the foliage, but it may prove necessary to arrange netting across doors and ventilators, as birds will eat the fruit whenever they get the chance.

When grown for commercial fruit production, the common *P. edulis* is usually spaced at 3–6m (10–20ft), the vines trained on trellises of post and wire set up in rows 2–3m (6–10ft) apart. Two leading shoots are allowed to grow from each young plant, and these are kept free of sideshoots until they have reached the wires and can be trained and tied in. Little pruning is called for after this stage, beyond occasionally pinching out the growing tips to encourage bushiness where growth is sparse, but if it should become necessary the trails and foliage can be cut hard back during the winter without harm. No crop is harvested the first year, but thereafter in the subtropics there are two main fruiting seasons per year. The fruit does not ripen satisfactorily after picking, and must normally be harvested only when

The fruit of the granadilla, *Passiflora edulis*, takes from four to six weeks after pollination to ripen

ready to use – sometimes even being allowed to drop to the ground and be gathered up routinely every other day. Ripe fruit furthermore will keep no longer than a week or two without deterioration, and these factors combine in limiting the production of passion fruit for temperate markets. When grown for fruit production, granadilla vines tend to lose their vigour after a few seasons and, in Australia and South Africa at least, are normally replaced after five or six years.

Light soils are best for the passion fruit, made fibrous with the addition of a little stable or farmyard manure, and heavy or poorly drained soils should be avoided. Inorganic fertilisers should be given sparingly, the chief benefit deriving from potassium in the form of sulphate of potash. For cultivation under glass the roots must be restricted if granadillas are to crop well, otherwise vegetative growth tends to become rampant at the expense of flowers and fruit. A concreted bed usually gives the best results, boxed about 30cm (1ft) deep and 1m (3ft) wide, but the drainage must be perfect, and a lumpy, turfy, sandy loam is ideal.

Tropical insects and also humming birds, in their native habitats, assist in achieving the cross-pollination necessary for heavy cropping. The biggest and best-flavoured fruits result from hand pollination, brushing each flower thoroughly in turn, as the quantity of pollen deposited on the stigma apparently influences both set and subsequent size. In the case of the typical purple granadilla, the flowers open at dawn and close at noon, and in the case of the yellow *P. e. flavicarpa*, they open at noon and close at sundown; in both cases they are most receptive to pollination midway through their daily opening period, the optimum stage corresponding with a characteristic curving of the prominent styles. Heavy moisture prevents effective pollination, apparently causing the pollen grains to burst prematurely, and the weather ideally should be dry at flowering time. Rainfall, however, is beneficial as soon as the pollination process is complete. Under glass, spraying or syringing should be strictly avoided until the fruit has set, but thereafter it is beneficial and should be practised frequently. The time lapse between pollination and ripening of the first fruit averages four to six weeks.

Granadilla seeds, of course, are usually eaten along with the pulp. When required for sowing they can either first be washed clean or sown immediately with the pulp still adhering, whichever is convenient, using standard seed-sowing compost in a shaded seedbox, set over bottom heat of 24°C (75°F). In any event they should be sown within a few months of the fruit ripening, as they do not remain viable for very long. As soon as the first true leaves have developed the seedlings should be potted up into small paper pots, using a lightly fertilised compost of 2 parts sharp sand, 1 part loam and 1 part sphagnum peat, plus a fair admixture of bonemeal. They can be planted into their permanent beds by the time they are 30cm (1ft) high, which, in the subtropics, will be about three or four months after sowing. For vegetative propagation of the granadillas, see passion flower, page 85.

Guava
Psidium guajava (Myrtaceae)

An attractive shrub with thin, flaking bark mottled in shades of green, brown and red, rounded light green leaves, and 3cm (1in) wide white flowers in midsummer, bunched rather inconspicuously two or three together in the leaf axils, followed by the fruit of varying colour, size and shape; this native of tropical America has spread extensively throughout the tropics of the world during the past few centuries.

As an introduced plant, guava naturalises itself rapidly, the seeds carried quite naturally by humans and birds. For the most part it has been welcomed as a useful addition to the local diet and native flora of the countries in which it has become established, but as rural populations of the Third World understandably prefer not to cut down providential fruit trees, it has attained the status of a serious weed in some regions – particularly where it has invaded cattle lands, and crowds out the grass with its suckering, thicket-forming habit. In the Philippines and some other Far Eastern islands it has run riot, and in Fiji it has been declared a noxious weed. I have seen the same process under way in Mozambique – the erstwhile Portuguese East Africa – for the early Portuguese travellers were greatly instrumental in the spread of guavas and other tropical foodplants. In one hilly region there it has spread radially from an old mission station, until in places between the 600m (2,000ft) and 1,500m (5,000ft) contours it has become the commonest woody species, everywhere crowding out the native bush on open hilltops, checked only by the almost impenetrable patches of tall evergreen rainforest which occupies the more sheltered valleys.

Under cultivation the guava is well behaved, and the fruit is grown commercially in some parts of Florida and California. Beyond the subtropical areas it makes a neat shrub for tub-planting. It will not tolerate frost but, this apart, the guava will withstand a wide range of tough climatic conditions and extreme soil types, surviving drought and flood, high temperatures and exposure to strong winds.

Probably the best dessert variety for size and flavour is the yellow-skinned *Psidium guajava pyriferum*, the fruits usually lemon- rather than pear-shaped despite the sub-specific name, 12cm (5in) long, with a sweet and juicy pink pulp; but plants grown from seed vary greatly in the quality of their fruit. The bushes start fruiting at a very early age, often only two years after planting out, come into full bearing at eight years and remain fruitful for thirty years or more. Planted in open ground they respond well to any fertiliser which is high in nitrogen, and a mulch of stable manure keeps the foliage green and lush. Occasionally guava bushes become chlorotic and benefit from an application of magnesium sulphate, or a spray of foliar feed which contains trace elements. A suitable general fertiliser might consist of 2 parts by weight ammonium sulphate to 1 part potassium sulphate, applied at the rate of 90g/sq m (3oz/sq yd) in the spring. Guavas are usually self-fertile; the fruits ripen about five months after flowering, and can be picked and stored before they are fully ripe. The yield is usually very heavy.

For tub-grown plants good drainage is important, and a sandy fibrous loam, to which has been added a proportion of leaf mould and dried farmyard manure, would make a suitable growing medium. A minimum temperature of 13°C (55°F) is needed in the winter, and 16°C (60°F) for healthy growth in the summer. If scion material is available from a good cultivar, seedlings can be patch budded – a more convenient method than the conventional shield or T bud, because of the thin, flaking nature of the bark. The stock seedling should have a neatly shaped patch of bark at least 2.5cm (1in) long removed in the early spring, at a point on the stem about 30cm (1ft) above the ground, and a similar but slightly larger piece is removed from the scion, embodying the dormant bud and including as a rule a thin sliver of wood backing the bud itself. This scion patch is trimmed to size, and sealed in place with transparent polythene tape, leaving the bud exposed. If still green and healthy one month after budding the patch can be released from the tape, and the top of the stock plant should be cut off as soon as the new shoot is about 2–3cm (1in) long.

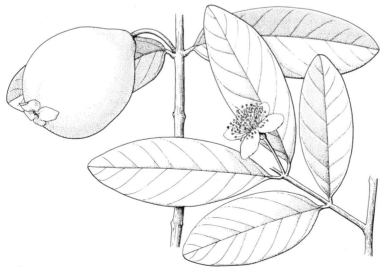

Psidium guajava

129

Guava seeds retain their viability only for about a year, and germination takes place two or three weeks after sowing. Seedboxes should contain a compost consisting of 2 parts fibrous loam, 1 part sphagnum peat, and 1 part sharp sand, embodying a little superphosphate and a sprinkling of ground limestone; the seeds are covered to a depth of 1cm (¼–½in), and the box set over bottom heat of 27°C (80°F). When potting up the seedlings it should be noted that guavas are shallow-rooted plants, and pots of the type used for rhododendrons are eminently suitable. Like rhododendrons, the roots of adult plants should not be buried deeply, and the large tubs used for permanent planting can also be fairly shallow.

Cuttings can also be used for propagation, taking shoot tips immediately below a node as the wood is firming in the late summer, placing them in a compost of pure sharp sand plus a little super-phosphate, over bottom heat of 24°C (75°F). Suckers often provide a convenient mode of increase, and these can be induced on plants grow-ing outdoors by cutting the lateral roots 70–80cm (2½ft) away from the main stem; but if taken from a budded cultivar their fruit will not, of course, be true to type, and they must themselves be budded with scion wood taken from the top growth.

Papaw
Papaya; pawpaw
Carica papaya (Caricaceae)
also Hardy pawpaw
Asimina triloba (Annonaceae)

A tree of exotic appearance, the tropical papaw is typically straight and thick-stemmed, with large, evergreen, ornately lobed leaves up to 75cm (2½ft) across, closely arranged around the trunk. The fruits also hang close to the stem, and measure from 10–30cm (4–12in) long, with a yellowish rind when ripe and bright orange flesh.

Papaya will not tolerate frost; its normal range of successful, productive growth lies in the hot lowlands between latitudes 32°N and S, and a low temperature results in poorly flavoured fruit. It does best in a rich, loamy soil, will not endure waterlogging, and needs a site in full sunshine. A native of tropical America, it thrives in Florida south of Palm Beach, where the fruits ripen between midwinter and midsummer, and reaches the extreme northerly limit of its productive range in southern California, although it is grown there sparingly

130

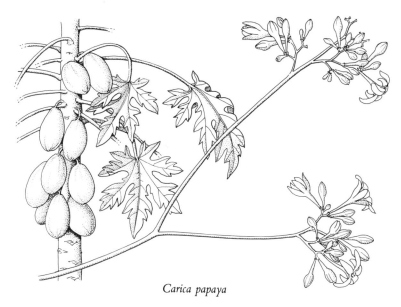

Carica papaya

within the hot interior valleys. As an ornamental, non-fruiting plant its outdoor range is considerably wider, and in many warm countries throughout the world it is used as a tropical bedder, often grown in a tub and plunged outside in the summer to form the central dot plant in a floral scheme. To produce fruit under glass it needs a minimum winter temperature of 18°C (65°F), and a minimum summer growing temperature of 21°C (70°F).

In tropical orchards, papaws are planted about 2.5m (8–9ft) apart, and after watering well are usually mulched with manure and kept shaded, either with some kind of temporary cover, or by low vegetation – often with bracken (an amazingly cosmopolitan plant). Under these conditions the trees come into bearing when only about one year old, sometimes as young as nine months, and can be expected to live about twenty-five years; fruit production declines as they age, however, and they are usually replaced before this. Depending on the age and condition of the tree, between thirty and one hundred and fifty large fruits are normally gathered per tree per annum. Papaws are harvested as soon as the skin starts to turn yellow, the stem cut cleanly with secateurs or a knife, and they will have ripened fully within four or five days after picking.

As a rule *Carica papaya* is dioecious, needing both male and female plants for fruit production, but a few self-fertile, hermaphroditic cultivars exist, notably Solo, a dependable variety with pear-shaped fruits, uniformly 15cm (6in) long and about 500g (1 lb) in weight.

131

Papaws grown from seed can be expected to inherit the typical dioecious characteristics, and for orchard planting it is necessary to ascertain the sex of each tree. This can be done as soon as flowering commences at the age of about eight months, when the small greenish-yellow male flowers can be readily distinguished from the females, which are noticeably larger and thicker. One male plant spaced in the centre of a group of six female trees will pollinate them all successfully. When grown as a greenhouse plant in temperate zones it is necessary to be equally selective when the flowers appear soon after midsummer, and if the plants are dioecious the process of fertilisation should be assisted by brushing the female flowers with pollen.

For propagation, seeds should be cleaned of the gelatinous pulp which clings to them, by scouring in sand. Papaw seeds remain viable for two or three years, but if very old or dry should be soaked in water for a few days before sowing, setting them 1cm (½in) deep in a compost consisting of 2 parts loam, 2 parts sharp sand and 1 part sphagnum peat, with the addition of a little superphosphate and a sprinkling of hoof and horn meal, shading the box or pots lightly and setting over bottom heat of 18°C (65°F). Papaw seedlings are often attacked by damping-off fungi, and to counter this should be watered with a solution of Cheshunt compound – 11 parts fresh ammonium carbonate to 2 parts powdered copper sulphate – dissolved at the rate of 3g/litre (1 oz of dry mixture to 2 gallons water), a dilution which can be repeated safely for every watering until potting up. Germination takes two or three weeks, and while the seedlings are still in the cotyledon stage they should be given full sun. The admixture of superphosphate and hoof and horn meal, plus their own generous nutrient supply, will keep them growing healthily in the seed-sowing compost for a month or so, after which they should be potted up into a rich loam to which a good proportion of well-decayed manure has been added. In the tropics papaw plants are ready for planting out after standing a further three weeks or a month, by which time they will be 15–20cm (6–8in) high.

Cuttings may also be used for propagation, taking ripe shoots with their leaves attached and setting them in a closed frame over bottom heat of 27°C (80°F). To encourage the growth of new shoots suitable for cuttings or grafting scions, the parent plant may be topped during the winter, but there is a risk in this as the stem is hollow and spongy, and liable to bleed sap profusely the following spring.

Asimina triloba, also known as the papaw, or hardy pawpaw, is a

(*Above*) Once well established, the coral tree, *Erythrina crista-galli*, develops a solid, hardy bole from which new shoots grow rapidly in the spring (*Chapter 5*). (*Below*) Even in quite cold areas the beautiful crape myrtle, *Lagerstroemia indica*, can make a vigorous, free flowering bush for sheltered gardens (*Chapter 5*)

North American deciduous shrub or small tree of the custard apple family, and a hardy native of rich woodland soil from western New York State to Nebraska, and southwards to Florida and Texas. It grows from 3m (10ft) to as much as 12m (40ft) in the wild, with large leaves and reddish-maroon 5cm (2in) nodding flowers in May and June, followed by purplish banana-shaped fruit up to 13cm (5in) long, ripening pulpy and sweet in the autumn. In cultivation, *A. triloba* is usually propagated from seed, and makes a rather gaunt garden shrub, quite hardy in the British Isles.

Pineapple
Ananas comosus (Bromeliaceae)

A native of tropical America but now of widespread cultivation throughout the tropics of the world, pineapples were once treasured and cosseted as luxury fruit in the great gardens of many temperate countries – until low-priced imports became general, and the custom died out. Since then, however, there has been a renewal of interest in growing the pineapple under glass as an interesting plant, with the bonus of an occasional fruit. The variegated form is particularly attractive in appearance and fairly free-fruiting; it tolerates typical houseplant conditions better than the green kinds though, like them, it must not be exposed to the cold draughts sometimes experienced on a windowsill during the winter.

Pineapples need a fairly dry soil, and the containers in which they are grown should be very well drained. The ideal compost would be based on a light, fibrous loam containing partially decomposed organic material to give it a rough, lumpy texture, perhaps embodying coarsely fibrous sedge peat and heather soil, composted bracken, and bonfire ash which includes small lumps of charcoal. Pot-grown plants can be brought to perfection in a compost consisting entirely of a mixture of peat and leaf mould. Fresh animal manure is not suitable, and feeding of any kind is rarely necessary; when a growth stimulant is required it should take the form of dilute liquid fertiliser.

Water is best given somewhat infrequently, in imitation of a monsoon climate – when needed it should take the form of a copious soaking, like a tropical downpour, preferably using water at a

Revelling in sunshine and heat, *Thunbergia grandiflora*, the Bengal trumpet vine, continues to open its magnificent flowers as long as the weather stays warm temperature of about 27°C (80°F). During winter and early spring water should be withheld altogether for two months. From late spring onwards through the summer, in addition to normal watering, plants should be sprayed with lukewarm water on sunny days, preferably during the afternoons with the ventilators closed, and containers of water should be stood around inside the greenhouse to create a humid atmosphere through evaporation. As the fruit develops and starts to ripen, water should again be withheld and the atmosphere allowed to become drier, for pineapples have a poor flavour if given too much moisture at this stage.

Adequate light is all-important, especially during early growth, and greenhouse plants should be allowed to stand as close to the glass as possible to keep them dwarf and sturdy. Only on very bright days should a little shade be given, preferably by moving the pots around and standing them beneath lightly foliaged plants. Pineapples should never be confined to a shaded greenhouse for the summer.

During winter the temperature should be maintained at a night-time minimum of 18°C (65°F), rising during the day to 21°C (70°F), a range which will rest the plants and keep them healthy without starting them into growth. At this season particularly, small pineapple plants can be carried indoors to double as houseplants whilst the regular ornamental subjects are being rested in the greenhouse. As the days lengthen, the temperature should be raised to a night minimum of 21°C (70°F), and if a heater can be sited beneath the bench, established plants benefit from warmth at their roots with an ambient temperature of around 30°C (86°F) maintained during the summer, with the greenhouse ventilated freely each morning.

Pineapples may be propagated from crowns, from gills, which are small growths arising just below the fruit, from stem cuttings, from suckers or from seed. Seed should be sown in shallow boxes or clay pans using a light, sandy compost, set beneath a polythene cover over bottom heat of 30–32°C (86–90°F). The seedlings should be potted into a light, peaty compost as soon as they are large enough to handle, and kept in a temperature of at least 29°C (85°F) until they have reached a substantial size. Crowns for propagation should ideally be removed with 2–3cm (1in) of the fruit attached; after the soft pulp has been scraped out, the rind is dusted with hormone rooting powder

Ananas comosus Variegatus, a variegated form of the pineapple, makes an attractive subject for greenhouse culture, with the bonus of an occasional fruit

and pressed into peaty compost topped with sharp sand, using a conveniently sized clay pot plunged in bottom heat of 32°C (90°F). Crowns, however, do not form such good, heavy-fruiting plants as do suckers, and they take longer to come into bearing. The severed stem of a plant which has fruited can be treated as a root cutting, laid flat in a box of sandy compost, buried to a depth of 3cm (1in) or slightly deeper, covered with a sheet of glass and placed over bottom heat of 27°C (80°F); shoots soon start to grow and take root independently, when they can be separated and potted singly. Gills may be rooted in the same way as crowns; but the large suckers which are formed at the fruiting stage develop rapidly and make the best plants. Suckers can be allowed to grow fairly large before their removal – twisting them off gently at any time between spring and autumn. After trimming the base smooth and removing the lowest leaves, they should be potted up singly in conveniently sized pots, usually between 13–20cm (5–8in) diameter. Good plants can be grown and made to bear fruit in a 13cm (5in) pot, and at no stage should they need one of larger diameter than 30cm (12in). Pineapples under glass are always easier to manage in pots than when planted out into a border; freedom to move the plants around the greenhouse means that full advantage can be taken of micro-climatic variations of shade and temperature.

5
SHRUBS AND TREES

Brunfelsia
Yesterday, today and tomorrow
Brunfelsia calycina eximia and others   (Solanaceae)

A native of Central America and Brazil, this bushy, upright, dark evergreen shrub can grow to 3m (10ft) in the tropics and subtropics, up to 2m (6ft) in other warm and fairly frost-free areas, and around 1m (3ft) as a not-too-demanding greenhouse plant within cool temperate zones. In the spring the bush becomes covered with sweetly scented 5cm (2in) flowers opening a deep purple with white centres, fading by the next morning to pale lilac, and by the following day to white – hence the name 'yesterday, today and tomorrow' – produced so profusely that flowers of all three colours are on view continuously for weeks on end. There is usually a second flush of flowers in the autumn which is not as a rule quite so prolific. The powerful jasmine scent of the flowers makes brunfelsia a popular choice for planting near a house door or beneath the windows, in countries where it will grow unprotected. Elsewhere it can be reared successfully as a tub-plant, and alternated to maximum effect between conservatory and patio, as the new flowers open between spring and midsummer. Fairly slow growing under these conditions, it is economical of greenhouse space, flowering as it does while still only around 40cm (16in) tall. Even out of the flowering season it is a handsome plant, the young foliage tinged with bronze and the mature leaves partially folded to expose a lighter green reverse.

Brunfelsia calycina eximia enjoys tropical conditions, but will tolerate an occasional light frost and has been known to survive winter temperatures as low as $-5°C$ (23°F). It grows happily in the major Australian cities; in North Island, New Zealand; the southern states and milder coastal regions of the USA; in East Africa and most of South Africa – it does well, for instance, at Pretoria, but not at nearby Johannesburg which, with its greater altitude, is a little too

The flowers of *Brunfelsia calycina eximia* open a deep purple, fading the next day to pale lilac and by the following morning to white – hence the name 'yesterday, today and tomorrow'

cold at times; its preference is for a copiously wet summer growing season followed by a moderately dry winter, though the terms 'summer' and 'winter' may be interchangeable in tropical countries according to local usage; in the coastal towns of southern Africa it grows to perfection, attains its maximum size and is often used as a vigorous hedge plant – a role to which it adapts very adequately.

Some of the brunfelsias flower throughout the winter if conducive growing conditions of temperature and moisture are maintained. The late-flowering sub-species *B. calycina macrantha* comes into flower soon after midsummer and will often continue blooming profusely until the following spring. The salver-shaped flowers, a rich violet-purple with white centres, often measure 8cm (3in) across the petals and form dense terminal clusters; a useful greenhouse shrub in temperate regions, but in order to stay in full bloom it needs a minimum winter temperature of 13°C (55°F), and a summer minimum of 15°C (60°F).

Other species include *B. undulata*, also an excellent shrub for the cool greenhouse, ultimately smaller overall than *B. calycina*, but with a vigorously upright stance that can be encouraged by careful clipping to adopt a formal spire. It has narrow dark green leaves with a paler green reverse, and terminal clusters of sweetly scented flowers which as a rule appear in the early autumn, pure white as they open, becoming yellow as they mature. Each individual flower takes the form of a narrow, curving, greenish-white tube some 9cm (3½in) long, widening into the wavy-margined cup of petals, some 5cm (2in) across. *B. americana* from the West Indies is known as the lady of the night, for the flowers are especially fragrant during the hours of darkness, they too opening a creamy white and changing to yellow as they mature. The long-tubed, salver-shaped flowers up to 5cm (2in) across are produced very freely in the late spring and early summer, and remain on the bush for a long period.

Brunfelsias enjoy a light but rich soil with a loamy base, made fibrous by the addition of leaf mould and sedge peat. During their growing season they benefit greatly from a weak solution of liquid manure added regularly to their water supply. Although flowering can be almost continuous with some of the species and sub-species while their growth remains active, as greenhouse plants they will live healthier, longer lives if given a definite resting period: the atmosphere as well as the compost should be kept moist during the growing season, but allowed to become dry and cool for the annual rest. As soon as container-grown plants finish flowering they should be repotted, or have some, at least, of their soil replaced, as they seldom thrive for many years if allowed to become root-bound. Trimming, if needed to keep their proportions neat, can be done at the same time. To start them into flower under glass they should be watered liberally, the surrounding atmosphere heated to at least 16°C (60°F), and the foliage syringed daily; but as soon as the first flowers open the syringing should stop and the atmosphere allowed to cool to 10°C (50°F) or less, thus in some measure simulating the cool 'summer' of Central America.

For propagation, sideshoots about 10cm (4in) long should be pulled

Brunfelsia americana

off with a heel of older wood which will need trimming, taking them soon after midsummer when the young growth is starting to mature and harden, and setting them in a compost consisting of 2 parts sharp sand to 1 part sphagnum peat, with the addition of a little superphosphate of lime and hoof and horn meal. The container should be set over bottom heat of 24°C (75°F) and covered with scored polythene, so as to permit the gradual entry of air as soon as new growth commences. When well rooted, they should be potted on into 10cm (4in) pots, using a fairly well-fertilised compost of 2 parts loam, 1 part sharp sand, 1 part leaf mould and 1 part sphagnum peat, and kept in an ambient temperature of at least 16°C (60°F) until they are well established and growing strongly.

Callistemon
Bottle brush
Callistemon citrinus; C. salignus and others   (Myrtaceae)

These handsome evergreen shrubs from Australia and Tasmania, with beautiful and quite characteristic prominently stamened flowers, are to be found in nature growing within areas of moderate or low rainfall (in the region of 45–85cm (18–34in) per annum), some of them conditioned to endure daytime temperatures reaching at least 38°C (100°F) over long periods during the summer. Those bottle brush species originating in the warmer Australian states are understandably tender and greatly subject to damage from even light frosts once their

142

growing season is under way; those from mountainous regions and from Tasmania are considerably more hardy. Many of the species prefer light soils; some occur naturally on rocky ridges, and a few are found in swampy regions. Under garden conditions, given a suitable climate, all respond well to a sunny site and a medium, well-drained soil.

Like *Eucalyptus niphophila*, the snow gum, Australian plants which originate in the mountains tend to be hardier under garden conditions than those species with a more southerly natural range, and the hardiest bottle brush species could well be *Callistemon pithyoides*, the so-called Alpine bottle brush, with small yellow flowers; *C. subulatus* is a cold-resistant, red-flowered mountain species from Victoria and New South Wales; next in the scale of hardiness is probably the yellow-flowered *C. salignus*, the willow bottle brush from Tasmania, also called pink-tip from the roseate tinge of its young foliage; the handsome red-flowered *C. citrinus* from Australia ranks next; the Tasmanian swamp bottle brush *C. paludosus* is fairly hardy; of medium hardiness are the crimson *C. rigidus* and *C. linearis*, both from New South Wales; two of the most tender of the genus are both from Western Australia – *C. phoeniceus* and the showiest of all the bottle brushes, *C. speciosus*, with deep scarlet flowers standing stiffly conspicuous amid narrow, sharp-pointed leaves in midsummer; finally *C. viminalis*, the weeping bottle brush, is if anything more tender than *C. speciosus*, and makes an untidy shrub unless trained to form a standard stem, when it weeps gracefully, the long scarlet flowers hanging profusely amongst leaves which are tinged pink when young, becoming dark green as they mature. A native of New South Wales, it has been cultivated very widely in the warmer parts of Australia as well as comparable areas in other parts of the world, such as East and low-veld South Africa. In such places it is often trained as a 6m (20ft) standard and planted as a street tree, having proved well able to endure poor, compacted soil and arid conditions. In the subtropics it is often the practice to trim these small trees every three or four years, thinning the foliage severely in order to increase the number of flowers and bring them into full view.

In northern temperate zones when given the protection of a cool greenhouse in which the winter temperature can be kept above 10°C (50°F), both *C. speciosus* and *C. viminalis* flower in the early spring. They cannot be expected to tolerate freezing weather outdoors, but several other callistemons will survive and thrive as garden plants in

Callistemon speciosus. One of the tenderest of the Australian bottle brushes, this beautiful shrub will not tolerate freezing weather

Callistemon citrinus Splendens is one of the hardiest of the Australian bottle brushes and stands exposure well

places where the winter temperature drops as low as $-12°C$ $(10°F)$, provided the soil is well drained, and some protection can be given from cold winds.

Near the south-west coast of England *C. citrinus* is reasonably hardy, both the wild species and its excellent improved variety Splendens, which has long-lasting flowers of brilliant scarlet. In some gardens it seems to survive particularly harsh winters better than *C. salignus*, but in most cases the reverse is true. In Devon a bush of *C. salignus* planted against a low south-facing wall but given no other winter protection endured a winter low of $-15°C$ $(5°F)$, where a neighbouring specimen of *C. citrinus* succumbed. Flowering is sometimes curtailed during the year following a harsh winter, but damage is mostly limited to frost-pruning in the early spring. As a general rule, although the habit of growth is affected in this way it is no bad thing for a small garden – the action of frost repeatedly nipping back the young shoots results in a far more compact and bushy specimen than any to be found in the subtropics, and subsequent flowering can be all the more attractive with the bush, in the case of *C. salignus*, taking on something of the appearance of a neatly rounded yellow-flowering American currant.

Several species of callistemon grow well in Ireland, and in more than one well-known sheltered woodland garden in the eastern and central districts, they seed themselves freely, or are sown *in situ* and allowed to run semi wild. In these circumstances *C. citrinus* is sometimes killed by frost during the early spring, but although *C. salignus* is occasionally cut back to the base during prolonged cold spells when the temperature may drop as low as $-14°C$ $(6°F)$, few of the latter have been killed completely. In the extreme south of Ireland in Bantry Bay, where the temperature rarely drops to freezing point, *C. citrinus* Splendens has proved tougher than *C. salignus*, and better able to withstand salt sea gales, for the area can be very bleak. The evidence here seems to point to the superiority of *C. citrinus* for tolerance to exposed sites, and to that of *C. salignus* for greater resistance to frost.

Seed provides the usual means of propagation for the bottle brushes. Seedboxes should be well crocked to provide ultra-sharp drainage, and filled to within 2–3cm (1in) of the top with a compost consisting of 2 parts sharp sand to 1 part sphagnum peat. To avoid losses from the damping-off fungi to which the seedlings are particularly prone, broken clinker or similarly free-draining material should be sieved to pinhead size and spread in a 1cm (⅜in) layer, the seed sown thinly on this and covered with a further light sprinkling of clean clinker. As an alternative to clinker, a sowing medium of pure sharp sand may be used, and the seedlings watered with a solution of Cheshunt compound (11 parts ammonium carbonate to 2 parts copper sulphate) dissolved at the rate of 3g/litre (½oz to 1 gallon water). In either case, the box should be stood over bottom heat of between 18–21°C (65–70°F) and kept well shaded, covering the compost with a sheet of paper until germination takes place. Seedlings should be pricked out into tubes or small, deep pots when they are between 3–5cm (1–2in) high, and when growing well should be stood in full sunlight in a sheltered place outdoors if possible until the autumn, over-wintering them in a cool greenhouse or frost-free frame.

As an alternative to seed, and necessarily in the case of cultivars such as *C. citrinus* Splendens, midsummer cuttings may be taken of semi-mature sideshoots 8–10cm (3–4in) long, pulled off with a heel. A 15cm (6in) pot is convenient to use, the compost in the lower half to comprise equal parts of sand and peat, but with the cuttings set around the edge of the pot in a layer of pure sharp sand. Callistemon plants which have been grown from cuttings are said to flower at an earlier age than those raised from seed. Other recommendations for vegetative propagation include short nodal cuttings taken at the point where the past and the current years' growths combine, and also soft tip cuttings taken at any time during the growing season, which will root fairly readily under intermittent mist.

Cassia
Golden shower; autumn cassia
Cassia corymbosa and others   (Leguminosae)

Small annual and perennial cassias may be found growing wild throughout the eastern USA as far north as the Canadian border, and some of these plants are very attractive, but the most beautiful members of the genus, the trees and tall flowering shrubs, are mainly natives of the tropical zones of the world.

Cassia didymobotrya is a native of tropical Africa, and a favourite garden tree in the warmer regions of South Africa and Australia, often known as the peanut butter cassia from the odd scent of its beautiful flowers, the petals a bright buttercup yellow in contrast with the dark sepia of the bud bracts which remain to clothe the tips of the flowering spikes very distinctively before the topmost flowers open. Commercial

146

senna pods are produced by *C. fistula*, a native of the drier parts of India and Ceylon, also very useful as an ornamental tree and variously known as the golden shower, the pudding-pipe tree or the Indian laburnum; the 60cm (2ft) long black pods follow a bright display of 30cm (1ft) long pendulous yellow sprays. *C. grandis* is the so-called pink shower from tropical America, a fairly tall tree; and another sizable cassia with pink flowers is *C. javanica* from Indonesia, an equatorial mountain species that does well in other mountainous tropical zones and such subtropical areas as experience mild frosts, often to be seen as an effective street tree in many warm countries of the world.

One tropical member of the genus adaptable to cool temperate regions, where it is able to survive not only in a conservatory or cold greenhouse but also outdoors in the angle of a sunny wall, is *C. corymbosa* from South America, a popular garden plant in South Africa where it is known as the autumn cassia from its season of flowering. In north temperate zones the large clusters of rich golden-yellow flowers in the form of 3cm (1in) wide cups with curving stamens and prominent crimson-brown anthers, usually appear late in the summer and persist until the weather starts to turn cold; specimens under glass may still be in full flower well into autumn, the long-stemmed pendulous clusters bright and summery amid the glossy dark green foliage. Autumn cassia as a rule takes the form of a 2.5m (8ft) shrub,

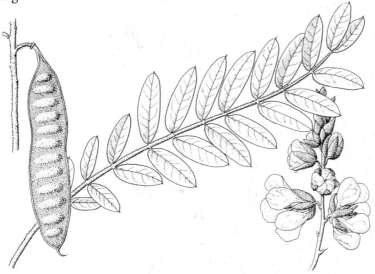

Cassia didymobotrya

147

and sends up long arching shoots annually from the base, in the manner of a leycesteria. A fast-growing plant spreading 2–3m (8–9ft) with ease, when necessary it tolerates severe trimming back as soon as flowering is finished – an operation best delayed until the spring in the case of plants growing outdoors in areas subject to frost for, although moderately tolerant, cold weather will sometimes kill the top growth to the ground. As a permanent resident of the greenhouse, however, continual cutting back of so vigorous a grower can result eventually in an unsightly thicket of shoots, and it is normal practice in ornamental horticulture to replace five-year-old autumn cassias with new stock. Outdoors in Britain, there is always the danger that a particularly cold winter will kill it, but in the English home counties at least it will survive against a south-facing wall through all except the most extreme winters, flowering freely in the late summer. South-east England is often the hottest part of Britain during the summer months; it is also frequently the coldest during the winter, however, and the most satisfactory autumn cassia plants in these places are those which have been established in large earthenware pots, annually plunged to soil level under glass for the winter, and outdoors for the summer. Grown in this way, the species makes an excellent dot-plant for exotic summer bedding displays. It can also be used as a tub-plant

Cassia corymbosa. In cool temperate regions the subtropical autumn cassia can be grown in a conservatory, or outside in the angle of a sunny wall.

to stand outdoors on the patio, though it has often been remarked that a cassia plant obliged to spend the winter standing in its container on cold greenhouse paving is far less likely to survive than one snugly plunged in a greenhouse border, or even set outside beneath a protective wall. In any case, the roots of pot or tub-grown plants must never be allowed to freeze inside their containers. Plants will survive provided the temperature does not drop below freezing, but a minimum of 7°C (45°F) is necessary to ensure healthy foliage and good flowering the following season.

In many parts of Greece and other countries near the Mediterranean Sea, the autumn cassia is often noticeable as a garden-escape plant which has become naturalised into the local flora – a maquis vegetation which includes other memorable flowering trees and shrubs: Judas tree, pomegranate, Spanish broom, bladder senna, oleander and Jerusalem sage. In the almost equally equable climate of south-western Ireland, at Tralee and Killarney where the strawberry tree grows wild, *C. corymbosa* flourishes as a wall shrub, and in one beautiful garden it enjoys the company of hoheria, eucalyptus, pittosporum, drimys, coprosma and cantua, in an exotic collection of shrubs from the southern hemisphere.

Cassia seeds germinate fairly readily in a compost of 1 part loam, 1 part sharp sand and 1 part sphagnum peat, to which has been added a little superphosphate of lime, the seedbox set over bottom heat around 24°C (75°F). Alternatively, cuttings may be taken in the early summer, using sideshoots as they reach a length of about 15cm (6in), or nodal cuttings selected from the top third of the current year's leading shoots around midsummer, as they are ripening at the base. These should be set in a compost consisting of 2 parts sharp sand to 1 part sphagnum peat, keeping the container uncovered, and using bottom heat of about 24°C (75°F).

Cestrum
Bastard jasmine; night-flowering jessamine
Cestrum spp.   (Solanaceae)

These beautiful, fragrant shrubs are natives of the West Indies and of tropical and subtropical America, reaching the limit of their natural range where near-freezing temperatures normally occur. The deciduous species especially are able to tolerate cold weather once they have become dormant for the winter, but even they are often set back severely by late spring frosts, and can be cut to the ground or even killed outright if the first frosts of autumn are unusually early or particularly hard.

Among the species most often seen outdoors in temperate climates, *C. aurantiacum* is one of the best, a rambling sub-evergreen shrub reaching a height of some 2.5m (8ft), with rounded leaves and 2cm (¾in) bright orange flowers in the summer – a plant best suited for sunny sites in mild areas. Like the other red- and orange-flowered species it is sometimes given separate generic status and classified as *Habrothamnus*. This showy Guatemalan native is seldom out of flower when grown under glass, but it has also been grown very frequently out of doors in sheltered positions, in the eastern coastal states of the USA and in England; a sunny wall provides the safest site, but it is always prudent to take the annual precaution of over-wintering a few cuttings in case replacements should be required.

Cestrum purpureum (syn. *C. elegans*) is considered hardy in southern British areas such as Cornwall and the Isle of Wight, but even there it grows better with the protection of a sunny wall. A tall, rambling, evergreen plant with pointed downy leaves, it produces dense clusters of 2–3cm (1in) reddish-purple flowers in summer and autumn, lasting sometimes until the first frosts. On an experimental basis, *C. purpureum* has been planted outdoors in northern England at the University of Leeds and, predictably, the results confirm that while the species will survive mild winters in the district – though even then it may be cut to the ground – it cannot be relied upon for normal garden use. In the greenhouse, however, it needs no more than minimal frost exclusion, and good container-grown specimens are often to be found in unheated conservatories. A vigorous plant, when grown in this way *C. purpureum* needs potting on regularly until a 30cm (12in) size pot is reached, where it should stay and be sustained by feeding. Old plants tend to become woody and unattractive, and it is then better to propagate them afresh. In a conservatory or garden room receiving some warmth, *C. purpureum* can be trained to make a beautiful archway, as much as 3m (10ft) high; summer temperatures should not be excessive, but when grown permanently in this style a winter minimum of 7°C (45°F) is advisable. Both feeding and watering should be generous during the summer months; the pots must be kept on the dry side for the winter, but the leaves will drop if the compost becomes too dry.

C. fasciculatum from Mexico is highly acclaimed as a flowering

(*Top left*) The crane flower, *Strelitzia reginae*, is seldom without a few blooms (*Chapter 2*). (*Top right*) The spectacular *Bougainvillea glabra* can be encouraged to flower several times during the year (*Chapter 3*). (*Centre left*) A glorious profusion of flowers on the marmalade bush, *Streptosolen jamesonii* (*Chapter 3*). (*Centre right*) In cool temperate zones the tropical climbing shrub *Clerodendrum thomsonae* is strictly a greenhouse resident (*Chapter 5*). (*Bottom*) A double yellow variety of *Hibiscus rosa-sinensis* (*Chapter 5*)

greenhouse shrub in Britain and the USA. A slenderly branched evergreen attaining 2–2.5m (6–8ft), both the stems and the 12cm (5in) leaves are densely covered with soft purplish hairs which give the foliage an overall purple appearance, and the 2cm (¾in) bright rosy-carmine tubular flowers are crowded in clusters some 8cm (3in) across. If grown outdoors in mild parts of the temperate regions, it usually survives and flowers well, but loses its evergreen character and sheds its leaves during the winter.

C. nocturnum, the night-blooming jessamine from the West Indies, is a 3.5m (12ft) evergreen which produces its 3cm (1in) greenish-yellow flowers in the late summer and autumn. They are not of striking appearance as they remain closed during daylight hours, but give off a delicious fragrance when they open at night. Like many of the genus, though the flowers are so fragrant, the foliage emits a foul smell if it is bruised or crushed, so that the plant is best not sited alongside a path, or near a doorway where clothing might brush against it.

The hybrid *Cestrum* Newellii is the most richly coloured of all the bastard jasmines with its bright crimson flowers, which appear in the latter part of summer and persist until the autumn, and it also is delicately perfumed, the fragrance particularly noticeable at night. This exquisite evergreen, of unknown parentage and not to be found in the wild, is quite as hardy as *C. purpureum*, and makes a fine wall plant for areas where the winter temperatures are unlikely to drop below −5°C (23°F). In Cornwall, England, it has been measured on a house wall at 4.5m (15ft), and produces a mass of scarlet flowers every year; during particularly harsh winters the top few feet die back, but the plant recovers and shoots again in the spring. Comparable plants are to be found against walls in coastal Ireland, and one notable 3.5m (12ft) specimen occupies a sheltered courtyard in the Shetland Isles, north of 60° N latitude – surely the most northerly cestrum to survive outdoor life. *C.* Newellii is able to grow in the shade, but such growth will be barren, and if given a warm but sunless position in the angle of a wall it will postpone flowering until the topmost growth has reached the sunshine.

C. parqui is hardier than most other bastard jasmines, a 3.5m (12ft) shrub with tapering 12cm (5in) deciduous leaves, well able to grow

Telopea truncata, the Tasmanian waratah, is moderately hardy and will thrive in a woodland garden with moist, peaty soil

153

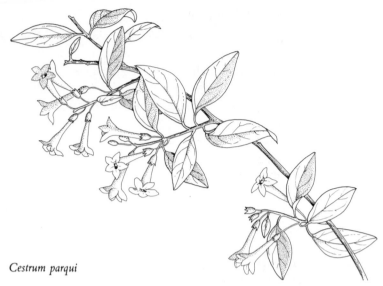

Cestrum parqui

outdoors in London given a reasonably sheltered nook, where it produces enormous clusters of 3cm (1in) long greenish-yellow flowers in mid and late summer, not particularly showy but very sweetly scented. That very skilled gardener, Margery Fish, prized this cestrum in her garden, not only for its fragrance but for the subtle combination of greenish yellow with the blue foliage of a nearby *Eucalyptus gunnii*.

In the greenhouse at Wisley Gardens, staff of the Royal Horticultural Society find they need to cut back some of their cestrums

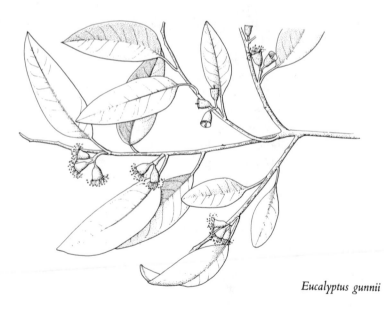

Eucalyptus gunnii

154

quite severely in late winter to keep them within bounds. Where there is room enough to allow them to grow unchecked they are in flower there almost continuously, given a minimum winter temperature of 10–13°C (50–55°F), but *C. parqui, C. aurantiacum, C. purpureum* and *C. Newellii* are all somewhat over-robust with such generous treatment under glass, and need constant attention to staking, tying and pruning back.

Propagation may be achieved quite readily with cuttings. In the case of *C. parqui*, soft tip cuttings should be taken in the late spring and early summer, cut cleanly below a node some 5cm (2in) below the shoot tip. After removing the lower leaves they should be inserted 2–3cm (1in) deep in a compost consisting of 2 parts sharp sand and 1 part sphagnum peat, surfaced with fine silver sand, and set fairly close together in deep boxes, over bottom heat of 27°C (80°F). The other species give better results using sideshoots, which are readily obtainable in the late spring as they reach a length of about 5–8cm (2–3in), and pulled off with a heel of older wood, just as they are firming at the base, using the same type of compost and bottom heat as for tip cuttings. They will usually have rooted well within three weeks, and should be potted on before they become too crowded in the boxes. As a precaution against damping off, as soon as fresh top growth appears the cuttings should be watered once or twice with a solution of Cheshunt compound, made with 11 parts ammonium carbonate to 2 parts copper sulphate, mixed and stored dry until required, and then diluted at the rate of 3g/litre (1oz to every two gallons of water).

Chaste Tree
Agnus castus; Abraham's balm; monk's pepper tree
Vitex agnus-castus   (Verbenaceae)

This attractive shrub is a native of southern Europe and western Asia; a somewhat spreading plant reaching a height of around 3m (10ft), with multiple stems arching from the base to 2.5m (8ft) or so, with deciduous, hairy, grey, aromatic, compound leaves formed of leaflets reminiscent of the spread fingers of a hand, and upright clusters of fragrant, violet-lavender flowers which appear in the early autumn. Later in the autumn the foliage takes on rich purple-bronze tints before falling for the winter – a valuable attribute in Mediterranean countries which often lack the glorious autumn coloration common to

Vitex agnus-castus alba, the white flowered form of the chaste tree, flowers in the autumn only after a long, warm summer. In dull climates it needs a sloping, sunny site

northern landscapes. There are also white-flowered forms, such as the wild variety *Vitex agnus-castus alba*, and the selected cultivar Silver Spire, which has larger and particularly fragrant clusters of individually quite tiny flowers.

In its native habitat the chaste tree is able to grow well and flower prolifically in poor, dry, stony soil, and this has been the experience also of gardeners in other warm countries to which it has been introduced; but the quality of flowering freely depends on the summer weather, and in cool, dull climates a poor soil cannot provide a suitable growing medium. A good soil compensates to some extent for a regional lack of sunlight and warmth, and the vegetative growth at least will be strong and luxuriant in consequence; conversely, a summer which is short and overcast – as often seems to be the case in Britain – means that the flower buds open only to half their extent; during a long, hot summer the trusses expand sometimes into 20cm (8in) spikes on the branch tips, the two-lipped lavender flowers each about 1cm (½in) long. In climates such as those of New York and the New England states, where the summers tend to be hot and the winters cold, the chaste tree flowers as well every late summer and autumn as the somewhat similar *Buddleia davidii* does in the UK; but the British summer is rarely long or hot enough – that of 1976 was a notable exception – to ripen the wood and open the flowers of most specimens growing in the open. Not merely flowering capacity, but the sheer survival of the species seems to depend in some measure on the comparatively extreme continental type of climate as opposed to the mildly temperate Atlantic influence; although the chaste tree shows great resistance to winter frosts once the season's growth has become hardened, if the summer weather has not been positive enough to ripen the wood the whole plant is liable to succumb. Where local climatic conditions encourage vigorous growth, the chaste tree may be treated in the same way as *Buddleia davidii* and cut back to within a short distance of the ground every spring. In countries such as Britain, where annual growth is less vigorous, it is safer to limit pruning to the removal of the old flowering shoots in the early spring.

The circumstance next best to a reliably long and hot summer is a site on a warm, sun-facing bank or wall, ideally on sloping ground to enable frosty air to roll safely away, and with a well-drained soil. A south-east aspect will often give the best results in Britain, for the micro-climate then tends towards the continental extremes, without involving over-exposure to the more bitter north-easterly winds which sometimes sweep across from Siberia. On the terrace of Powis Castle in central Wales the chaste tree grows well with a south-easterly aspect, in company with other southern continental trees, and comes into flower beautifully most years.

In ancient times the seeds of the chaste tree were considered a potent anti-aphrodisiac, and used in various traditional ways to ensure chastity. As seed is not usually set unless the climate is completely suitable, propagation is usually achieved vegetatively, conveniently with cuttings of half-mature heeled sideshoots 5–7cm (2–3in) long, taken just as they are firming towards the base. This optimum stage of ripeness corresponds with the opening of the earliest flower buds. The cuttings should be inserted quite close together in 15cm (6in) pots, using a compost of 2 parts coarse sand, 1 part loam and 1 part sphagnum peat, embodying a little superphosphate of lime, and surfaced with about 1cm (½in) of fine, sharp sand which is allowed to trickle to the base of the cuttings as they are inserted. They should be watered well into place and the pots plunged in sand over bottom heat of about 18°C (65°F), if possible covering the frame with scored polythene. As soon as new growth appears air should be admitted, increasing exposure gradually until the rooted cuttings are quite hardened off when they should be potted singly into a loamy and fairly well-fertilised compost.

Clerodendrum
Glory tree
Clerodendrum bungei; C. thomsonae; C. trichotomum; C. ugandense and others   (Verbenaceae)

Many of the shrubs and some of the climbers included in this huge genus are valuable in the garden for their often sweetly scented late summer and autumn flowers, and the brightly coloured bracted berries that follow. All are natives of warm countries, and for this reason not always easy to please when planted within the world's temperate zones.

The Far Eastern *Clerodendrum bungei* is one shrubby member of the genus that will grow outdoors in most cool regions, given a site sheltered from freezing winds and late spring frosts. Originating in a region of decisive seasonal rainfall, *C. bungei* appreciates a copious supply of water during the summer, followed by a period of dryness over winter. The foot of a wall quite often remains dry during rainy weather, and in particularly wet areas this type of shelter from the prevailing winter winds would be desirable. In the drier parts of southern England it should preferably be sited on a slope freely drained enough to allow cold air as well as surplus water to flow away from the stems, and often gives the best results near a west-facing wall where it will be shaded during the early part of the day. Rather heavy in foliage, with 20cm (8in) wide heart-shaped leaves (which emit a vile smell if crushed), it is conspicuous in the autumn when covered with flattish heads of fragrant, rosy red flowers. The wood remains soft in this mild climate, and consequently the top growth is often cut back to the ground by frost, but it grows away strongly in the spring, sending up dark, erect stems reaching about 2m (6ft), topped in the late summer and autumn with the flowers in clusters some 10–12cm (4–5in) across. It is a suckering plant, almost herbaceous in character, and whether damaged by frost or not sends up new shoots annually in the spring, sometimes to appear a couple of paces away from the parent plant.

C. trichotomum is somewhat hardier than *C. bungei*, less inclined to seek shelter from the wind, but thriving in full sunshine and unwilling to tolerate even partial shading. In the late summer and early autumn it produces loose 20cm (8–9in) sprays of lily-scented starry white flowers 3–4in (1½in) across, framed in reddish-brown calyces, followed as the petals fade by bright blue berries each backed with a star of crimson-maroon bracts in a striking combination of colours. An upright shrub or rather gaunt small tree with large glossy green leaves, this strong grower from China and Japan commonly reaches a height and proportional spread of some 4m (12ft), and specimens of 6m (20ft) or more are not unknown. The variety *C. trichotomum*

Clerodendrum bungei, a shrubby member of the glory tree genus, is cut back by frost but grows away strongly in the spring

The tender tropical shrub *Clerodendrum ugandense* is sometimes called the Oxford and Cambridge bush to describe the flower colour – two-tone in light and dark blue

fargesii usually flowers and fruits more freely than the type, and is sometimes classified as a distinct species.

The tropical *C. ugandense* is one of the most beautiful of the genus when in flower, a daintily slender shrub carrying clusters of long-stemmed 3cm (1in) blooms, each with four rounded light sky-blue petals and one long, tongue-like petal of dark blue, the flowers followed later in the season by black berries. In African gardens it is sometimes known as the 'Oxford and Cambridge bush' on account of its bicolor flowers, which are on show there for many months of the year. Within cool temperate zones it flowers in the autumn, and is far too tender to stand outside until the arrival of the first frosts of winter. As a greenhouse plant it needs a bare winter minimum temperature of 7°C (45°F), and even this is rather risky; 10°C (50°F) is preferable; in the temperate house at Wisley Gardens, where *C. ugandense* flowers from August to October – its blue flowers blending beautifully with the gold of autumn cassia and the white and pink flowers of the Australian bower plant – the temperature is maintained at a winter minimum of 13°C (55°F) and a summer minimum of 16°C (60°F), and the atmosphere kept reasonably moist during spring, summer and autumn, but fairly dry over winter. With wand-like stems radiating from the base, it grows rapidly into a straggling, willowy shrub, usually about 1.5m (5ft) tall. The Royal Horticultural Society recommends that *C. ugandense*, and all the tender shrubby clerodendrums when grown in containers as permanent greenhouse plants, should be cut back quite heavily after flowering, and kept on the dry side at least during the early part of winter, given enough heat to start them into growth early in the New Year, repotting and setting them over mild bottom heat as soon as new growth appears.

C. speciosissimum from Indonesia is a rapidly grown shrubby plant with vivid orange-red flowers in the autumn. It is often grown as a greenhouse annual but, despite its tropical origin, if kept permanently and treated as a shrubby perennial it will tolerate over-winter temperatures no higher than 10°C (50°F). The West African *C. splendens*, which has bright scarlet flowers, needs warmer treatment, with a winter minimum of 18°C (65°F), and at least 21°C (70°F) in the summer.

The climbing species also tend to need warmer growing conditions, like the evergreen *C. thomsonae* from tropical West Africa, a vigorous plant which carries loose sprays of spectacular crimson and white flowers in the summer. Strictly a greenhouse resident when cultivated in the colder temperate zones, it will not do well if winter temperatures are allowed to drop below 16°C (60°F). Though often grown in tubs or large pots, specimens soon become root-bound, and far better results are seen when the plants are grown permanently in a greenhouse border and trained up a wire frame to the roof ties. If this is not possible, container-grown specimens are best rested and made to lose their leaves for the winter, although evergreen by nature, by allowing the compost to dry out completely. Whether rested in this way or not, *C. thomsonae* does well in a compost consisting of equal parts loam and peat, with the addition of some sharp sand and leaf mould – a less rich growing medium than that needed by the shrubby species to flower well. After flowering the trails should be pruned back and the younger shoots shortened to about 8cm (3in). At the

Longwood Gardens in Pennsylvania, USA, *C. thomsonae* shares a tropical house with bananas and papaya standing amongst tea and coffee bushes, and competes with the bougainvilleas for brilliance of flower. It is also much admired in the greenhouse at Wye College, Kent, England, where its massed flowers are vivid in the early summer against the dark green foliage, and here also it is judged showy enough to share the startlingly exotic company of orange *Bougainvillea* × *buttiana*, and violet-blue *Petrea volubilis*.

Seed of the shrubby clerodendrums, if available, may be sown either as soon as it ripens in the late autumn or in the spring, using an evenly balanced sand/peat mixture in a seedbox standing over bottom heat of 24°C (75°F). A few species such as *C. speciosissimum* grow rapidly enough from seed to enable them to be used as greenhouse annuals or biennials – seed sown in the late summer resulting in plants which will be flowering by mid to late summer of the following year, and seed sown in the late winter or earliest spring producing plants ready to flower in the later summer or early autumn of the same year.

The shrubby species will grow readily from cuttings, and if plants are cut back annually in the autumn the resultant trimmings may conveniently be used, taking pieces of stem or sideshoots about 8–16cm (3–6in) long, and setting them in a sandy compost over bottom heat of 21°C (70°F). Cuttings of the climbing species will not take root so readily, and their container should be covered with scored polythene to keep the atmosphere close, stretching it a little to admit air as soon as growth commences.

Suckers provide a ready means of increase whenever they appear, particularly so in the case of *C. bungei*, which produces them rapidly and extensively. Young non-suckering clerodendrums will usually divide readily, and this should be done as soon as growth starts in the spring. *C. trichotomum* and its variety *C. t. fargesii* may conveniently be propagated from root cuttings, digging up the roots in early winter and, after washing them clean, cutting into 6–7cm (2½–3in) lengths, ideally about 5mm (³⁄₁₆in) thick, the bottom end given a slantwise cut for identification. These should be bundled and buried in sand, the right way up, in a sheltered place outdoors; in the early spring they should be lined out in open ground, setting them vertically about 15cm (6in) apart in sandy beds, burying the top of the roots not more than 1–2cm (½in) below the surface. In the case of the more tender clerodendrums, the root cuttings should be set in a peat/sand compost in deep boxes, over bottom heat of 16°C (60°F).

Crinodendron
Lantern tree
Crinodendron hookeranum; C. patagua (Elaeocarpaceae)

A handsome small evergreen tree or large shrub from Chile, *Crinodendron hookeranum*, if it likes its site, can attain 6m (20ft) or more and, following a reasonably warm summer the previous year, spring sees it spectacularly laden with pendant red lantern flowers. A mild winter and well-defined spring season are as important to flowering performance as the summer weather, for the flower buds have already been formed on the shoots long before they are due to open, and are therefore very vulnerable to damage by cold winds and late frosts. In mild areas during a favourable year, the long-stalked crimson lanterns can cover the foliage almost completely from spring until well into late summer. Though it will grow successfully in a soil of pH 7 if the climatic conditions are acceptable, crinodendron will not tolerate free lime at the roots, and thrives in the humus-rich, acid soil and the partial shade of a sheltered woodland garden.

In the Pacific coastal states of the USA, such as Oregon and Washington, *C. hookeranum* finds itself at home in a climate not dissimilar from that of its native habitat. Similarly in the oceanic climate of Ireland, western Scotland, and favoured districts of south-west England, the lantern tree is reasonably hardy – as at the woodland garden of Brodick Castle on the Isle of Arran at the mouth of Loch Fyne, where it forms enormous clumps of vegetation near a pond, in association with gunnera, astilbe and tall iris. Away from the woodland environment it can best be established in the shelter of a wall facing away both from the midday sun and from dry-cold continental winds – in Britain, a north-west aspect will be ideal, but in this situation it is all the more essential that the soil be similar to that found in a typical woodland – acid, moist and rich in humus. In Devon gardens it can be seen flourishing in the company of another Chilean flowering tree, *Embothrium lanceolatum*, and the two appreciate similar woodland conditions, but even here it is vulnerable to damage during exceptionally severe winters. It makes a good wall plant, even when under exposure to salt gales, not only in the English West Country where it can make a good companion for the banksian rose, and can tolerate temperatures as low as −12°C (10°F) without damage, but even in the cold east-facing northland of Northumbria where it has been seen growing well in the lee of a wall, though it is

Crinodendron hookeranum, the red lantern tree from Chile, will grow outdoors in mild temperate regions

cut back by the east winds whenever it shows its head above the shelter. While the species is quite hardy in the comparatively mild Western Scottish Isles, elsewhere in Scotland in the more extreme mainland climate, as in England, lantern trees in woodland conditions have often been badly damaged and even killed by harsh winter weather – the damage chiefly occurring when freezing gales blow from the east and strip them of shelter by uprooting the taller woodland trees standing in their path. Such freak conditions excepted, lantern trees are to be found growing healthily in woodland gardens throughout northern Britain: at Harrogate in Yorkshire the species thrives in an acid clay loam in the company of camellias and eucryphias.

Crinodendron patagua, another Chilean tree, which bears white lantern flowers in the late summer, is also sometimes planted in similar situations and, like the red lantern tree, makes a vigorously growing shrub or small tree. In the milder parts of coastal Ireland it is to be found here and there in woodland gardens, keeping company with other exotic plants from the southern hemisphere, tree ferns, acacias, griselinia and olearia, often sheltered by clumps of indigenous trees and shrubs, holly and scrubby oaks. Although the late flowering season of this species means that the buds escape severe winters unscathed, it is certainly no hardier overall than the red lantern tree, and the top growth is sometimes cut back by frost and cold winds.

Away from sheltered woodland areas, in the case of both wall-grown and free-standing plants of either species, winter protection can be provided in the form of sacking, wattle hurdles, or a frame of wire netting padded with bracken. Sacks – which should be free from any residues of lime or alkaline fertiliser – should be lagged around the stem and over the crotch of the crown, in such a way that even if the smaller branches are killed, the stem and main framework of large branches will survive to form the basis of a shapely tree.

Propagation of either species is normally effected with cuttings of semi-mature sideshoots taken with a heel in the late summer, when they reach a length of about 7cm (3in) and are just firming at the base. Fresh, clean growth should be chosen, and not the older, harder shoots stunted to a similar length, which are often to be found. Normally twelve or fifteen cuttings are inserted around the edge of a 15cm (6in) pot, as they seem to root better in company than when placed in individual pots. The propagating compost should consist of 3 parts sharp sand to 1 part sphagnum peat, surfaced with 1 cm (½in) of fine sand, of which a modicum is allowed to trickle to the base of each cutting as it is inserted. After watering well, the pot is placed in a closed frame over bottom heat of 24°C (75°F). The cuttings will normally have rooted within a month, after which time fresh air should be admitted by degrees. Once the case is fully open and fresh top growth has appeared, they should be potted singly in 7–8cm (3in) pots, using a moderately fertilised compost of 2 parts fibrous loam, 1 part sphagnum peat and 1 part sharp sand, and the pots stood in a cool greenhouse until the spring. As the new growing season starts, they should be moved into large-sized pots and stood outdoors in the shade for the summer months.

Daphne
Garland flower
Daphne cneorum; D. odora; D. petraea
also Mezereon
D. mezereum and others   (Thymelaeaceae)

The hardy daphnes have a reputation for being temperamental under cultivation, but these very beautiful fragrant-flowered shrubs have varied requirements: in the main they need an open, sunny position and a well-drained soil embodying adequate humus; many of them appreciate permanent moisture at the roots, but even in these cases the drainage must be very free; drought, on the other hand, can kill them readily. Most European and many Asian daphnes come from limestone areas, yet they seem to enjoy an acid soil; most grow better for the addition of peat or leaf mould in the soil, but very few seem to be lime haters, and of all the daphnes *D. retusa* is perhaps alone in shunning any trace of lime.

The deciduous *D. mezereum* is one of the easiest daphnes to grow, typically flowering in the late winter and early spring, with fragrant blooms of madder pink. There is also a white-flowered variety named *D. mezereum* Album, and a very vigorous cultivar known as Bowles White which has ivory-white flowers. The species seems especially to

Helleborus niger thrives in the limy soil favoured by the mezereon

thrive if allocated a somewhat cold, sticky, limy soil in the shade – the type of soil favoured by the Christmas rose *Helleborus niger*. In the sandy, acid soil of the Royal Horticultural Society Gardens at Wisley, though it flowers annually in March – in mild winters, as early as the second week in February – *D. mezereum* and its varieties do not do particularly well. The natural distribution of the species throughout Europe was probably originally confined to calcareous woods, though it seems unlikely that any truly wild mezereons remain today; they have been in cultivation for many centuries now and all, probably, found their way to gardens long ago.

When the fragrant flowers have finished their display, the equally ornamental scarlet fruits are decorative throughout the summer months, if the birds allow them to remain. Ripe mezereon seed is often eaten, mainly by blackbirds which at one time dispersed the plants by this means in nature; but all too often the fruits are taken prematurely in late spring and early summer before they ripen. In Britain, greenfinches are the worst offenders in this respect, eating them while the nutshell is still soft, unconcerned that both leaves and berries are reputedly poisonous. Once the local greenfinches have acquired the habit of feeding on mezereon during the brief period between the setting and ripening of the berries, they continue to do so year after year – an unfortunate habit as, apart from the aesthetic viewpoint, daphnes in general do not take easily from cuttings, and layering is time-consuming and somewhat cumbersome in practice. Mezereon in any case lives no longer than twenty or so years, and the seed is really needed for propagation; allowed to ripen on the bush, it will eventually germinate where it falls, though it may take a full year to do so.

Various measures have been tried to combat the destructive green-finches without actually harming these beautiful birds: the various scaring devices on the market prove effective often for little more than an hour or so before the birds become inured to them; spraying the buds with repellent chemicals seems ineffective as a deterrent: the traditional cat's cradles of black cotton stretched over the plants have not proved discouraging to greenfinches; and garden netting arranged to cover the plant usually allows the bird entry at ground level – or traps them by the feet; polythene bags to cap the fruiting sprays have been tried, but the birds can see through them and usually manage to tear the material. Birds are attracted to food by sight rather than scent, and apparently the only sure way to avoid their attention is to prevent

Daphne mezereum, the mezereon, is a woodland plant never found today in the wild. In the garden, its seeds need protection from birds

The sweetly scented purplish rose flowers of the garland flower, *Daphne odora*, appear in mid-winter, unperturbed by the weather

their seeing it. During experiments carried out at London, muslin bags tied round the spray clusters have proved most effective, especially if the muslin is thick enough to obscure the view. Once the crucial midsummer fruit-setting period has passed and the daphne nuts are hardening, greenfinches no longer feel the urge to take them, and the bags can safely be removed, leaving a fine ornamental feature for the remaining summer months.

The purplish-rose flowers of *Daphne odora* – sweetly scented, some claim of lemon – appear in midwinter, unperturbed by the weather. A low-growing evergreen in most temperate zones, it will grow reasonably well in the shade, and in warm climates needs at least partial shade from the hot afternoon sun, but it will not thrive in the tropics. Though the flowers are frost-hardy, severe frosts may defoliate established plants, forcing them to adopt a deciduous habit. Cold winds seem to be the worst hazard, but they nevertheless survive most cold temperate winters unharmed. Some of the attractively variegated forms such as the yellow-margined *D. odora* Aureovariegata are particularly prone to lose their leaves in the winter, but the plant almost always recovers, and many gardeners claim that this form is hardier than the type. In a woodland setting, *D. odora* enjoys hot summer seasons better than does *D. mezereum*, and it thrives over much of the southern USA: the Cypress Gardens of Charleston, South Carolina, are perfumed in midwinter by *D. odora* flowering in a majestic setting, later made colourful by the azaleas massed beneath silvery Spanish moss draped from the trees. In general, a fairly heavy, lime-free soil gives the best results, and an additional mulch of compost or leaf mould seems to be beneficial. Mulches often induce the lower branches to root, and this type of layering can easily be encouraged by setting the outer branches several inches deep in peaty compost, and leaving them for a year before severing and transplanting.

D. collina makes a dense, rounded, miniature bush producing a good show of rose-pink flowers in the spring and again in the autumn. *D. cneorum* is better known in the flower garden, though it is fairly difficult to establish successfully. Its roots appreciate a soil made rich with leaf mould, and a surface covering of stones to keep them moist, shaded and cool, although it succeeds both in limy soil, which it probably prefers, and in almost pure peat, where it makes a thick, prostrate mat. In hot, dry areas it appreciates an additional mulch of peat or leaf mould around its outer branches, and tends to do better in

169

Daphne cneorum

humid regions with fairly cold winters. The typical species and its cultivated variety Eximia both form clumps eventually reaching more than 1m (4ft) across, well covered midway through the spring with large, pink, sweetly scented flowers.

The miniature *D. petraea* in nature grows high on limestone cliffs in the Alps, and it also is difficult to establish in gardens. The deep rose-pink flowers appear towards the end of spring, strongly fragrant and waxy petalled, larger and more brilliant in the form known as *D. petraea* Grandiflora. Unwilling to root from cuttings and a slow grower, it is usually grafted in the spring on two-year-old seedlings of *D. mezereum*. Garden specimens sometimes lose their flower buds during the winter, and this may in part be due to some shortage of water experienced during the previous summer. This fine little daphne is often grown in a clay pan and confined to an Alpine house, but some of the best specimens are to be seen outdoors growing in tufa – the extremely soft limestone which is prepared by having 2–3cm (1in) holes drilled to take the plants, which are then firmed in with sand and watered well. Tufa is best soaked about once a fortnight during hot weather, and normally needs no watering at all between early autumn and the following spring.

D. blagayana is a prostrate species carrying dense terminal clusters of fragrant tubular creamy white flowers in the late spring. It favours a stony soil, and small stones can be placed around the branches to hold the rather untidily sprawling stems in place. In nature it is found in mountain forests in southern Europe within limestone regions, but lime seems not to be necessary for its successful cultivation in the garden, and it thrives equally in acid soils. The branches layer themselves readily if covered with soil or leaf mould. *D. laureola* is a European evergreen shrub – the wild British spurge laurel – sometimes used as a grafting stock for the evergreen daphnes. The rather similar *D. pontica* is a native of Turkey making a 1.5m (5ft) evergreen bush which thrives in woodland shade and a humus-rich but lime-free soil. Though not a beautiful shrub, its greenish-yellow spring flowers are valued for their powerful fragrance which is especially strong in the evenings. Other profuse spring flowerers include: *D.* × *neapolitana*, one of the easiest of daphnes, with sweetly scented pink flowers; *D. retusa*; *D. tangutica*; and the very beautiful *D.* × *burkwoodii*, also one of the easiest to grow, with sweetly scented pale pink flowers. *D. dauphinii* (syn. *D.* × *hybrida*) is good in heavy soil, and flowers in the late autumn and early winter.

Daphnes sometimes die suddenly for no apparent reason. Occasionally a virus disease may be responsible, but more usual causes are inappropriate cultural treatment and drought. Some species are in any case short-lived, a factor which is often overlooked. Daphnes which dislike lime often assume a chlorotic appearance, and may be restored to health with a course of iron sequestrene.

Daphne petraea Grandiflora. The exquisite pink flowers of this Alpine shrub appear towards the end of spring

Except in the case of a few species, daphne seed is notoriously reluctant to germinate. Commercially, both *D. laureola* – for use as rootstocks – and *D. mezereum* are frequently grown from seed. Following a time-honoured method the seed is gathered as soon as it ripens, washed free of pulp and stratified in slightly moist sand until very early spring, when it is sown in a compost usually consisting of 2 parts loam, 1 part sphagnum peat, 1 part coarse sand and some super-phosphate, with the propagating frame kept closed until the seedlings are about 5cm (2in) high. The following spring they are normally transplanted and lined out in sheltered outdoor beds. Even with traditional stratification, germination is somewhat erratic. Experiments with daphne seed have shown that 100 per cent germination can be procured consistently by using the following method: after sowing in pots or small trays in the normal way, the containers are placed in a temperature of 26°C (79°F) for four weeks; following this heat treatment the containers are transferred to a cold room and kept chilled at 2°C (36°F) for eight weeks; after this simulated change of seasons they are again warmed to 16°C (61°F) for a further eighteen days, at which point germination takes place.

D. collina, D. × *neapolitana, D. dauphinii* (syn. *D.* × *hybrida*), *D. odora, D. retusa* and *D. tangutica* can all be grown from cuttings taken in the late summer as the current season's shoots reach an average length between 7–10cm (3–4in). Cuttings are taken either immediately below a node or, if the length is convenient, with a heel of older wood. Procured at this stage, the lower half of the cutting should consist of fairly firm wood. In the case of nodal cuttings, rooting is improved if the base is split cleanly behind the bud to a depth of about 1cm (½in). Clay pots should be used in preference to pans or boxes, and the compost can comprise 2 parts sand to 1 part sphagnum peat, surfaced with a thin layer of coarse sand. Having been watered well, the pot should be covered with polythene or placed in a closed case, and given bottom heat of 18°C (65°F), shaded heavily, and examined regularly to ensure the compost is kept uniformly moist. After five weeks of this treatment they should be rooted well enough for potting on.

In the case of *D. cneorum* a different technique is necessary. Half-mature sideshoots about 5cm (2in) long should be taken with a heel in midsummer, and set close together in a compost of 2 parts sharp sand and 1 part sphagnum peat, surfaced with coarse sand. They should be watered well at first, but after covering the container closely with thin

The roots of the fragrant shrub *Daphne cneorum* appreciate a soil enriched with leaf mould and shaded with stones

polythene they should not be watered again except by immersion from below. The white-flowered form of *D. cneorum* is reluctant to strike roots, and this variety, like *D. blagayana*, can conveniently be layered in the late spring; in the case of *D. blagayana*, by the simple expedient of heaping soil over the prostrate branches which, after severing in the autumn, are lifted and potted the following spring; in the case of *D. cneorum* varieties, a generous quantity of peat and sharp sand should be worked around the current year's shoots, pegging or anchoring the layers with their bases at least 5cm (2in) below the soil surface – preferably giving the stem a slight twist so as to break the bark at the point to be rooted. They should have rooted strongly enough by the following spring to be lifted and lined out, setting them about 10cm (4in) apart in a moist but well-drained bed.

Cultivated varieties of *D. mezereum* may be grafted on two-year-old stock plants of the type species established in 5–7cm (2–3in) pots, using 5cm (2in) tips of the previous year's growth as scion material, and performing the operation as soon as the weather starts to warm up in the spring. In the case of *D. petraea* and its variety Grandiflora, two-year-old stock seedlings of *D. mezereum* are used. They should have been established the previous year in 5cm (2in) pots, and kept plunged outdoors in a sheltered spot until late winter, when the pots are lifted and placed in a cold frame. The optimum time to graft occurs just as the roots show their first signs of growth in the earliest spring. At this stage the stock plants should be headed back to within 4cm (1½in) of the soil, using a 7–15cm (3–6in) scion of the previous year's growth in a simple cleft graft. The frame should then be closed, and ideally given bottom heat of around 21°C (70°F). The union should be complete a month later, when air can be admitted by degrees, and they should be left a further two or three weeks with the frame open, before staking and tying.

Daphne genkwa

In the case of *D. genkwa*, root cuttings offer the most convenient means of propagation. Pieces of root should be cut into 1–2cm (½in) lengths and set singly in small pots, using a compost of 2 parts sphagnum peat, 1 part loam and 1 part sharp sand, plus a little super-phosphate of lime. The pieces should be placed horizontally and covered shallowly with pure sharp sand, before topping the pot up with a further shallow layer of compost and firming very gently. The pots are best plunged in an open case, watered well, and covered with paper until shoots appear; in cold weather bottom heat of about 15°C (60°F) is beneficial; rooting follows after a further two or three weeks, at which stage they may be transferred to a cool greenhouse and allowed to become well grown before potting on.

174

Erythrina
Coral tree

Erythrina crista-galli; E. lysistemon   (Leguminosae)

Erythrina crista-galli from Brazil is called the coral tree from its sweet pea-like waxy scarlet flowers which cluster in large terminal racemes, hanging gracefully on long stalks in the summer and autumn – a rangy, bushy, deciduous shrub with glaucous green leaflets, often grown as a greenhouse annual in cool countries, but reaching 3m (10ft) or more in the tropics and subtropics, where the flowers are often on display for six months of the year, often appearing in the spring at the same time as the leaves. In climates less amenable to rapid growth, the long, prickly shoots are sometimes best cut back each year to within a few inches of the heavy woody rootstock, an operation which can be carried out at any time between leaf fall in the autumn and the first new growth in the spring. This seemingly harsh treatment encourages the formation of flowers and results in a more shapely plant. The top-growth of outdoor plants in any case normally dies back in temperate climates subject to winter frost. Once the rootstock has become well established, the plant develops large pithy roots and a solid bole from which new shoots are able to grow rapidly in the spring.

In Britain and comparable climatic zones, *E. crista-galli* needs the encouragement of a nearby warm, sunny wall, and during the winter the rootstock should be covered with a thick layer of leaves, bracken or wood ash to prevent the pithy roots freezing. Once this solid crown is well developed the plant will certainly tolerate quite heavy surface frost, but losses do occur, and it is safer to exclude ground frost. Experiments have shown that a well-developed specimen growing outdoors in a region subject to very heavy frosting will survive a certain amount of freezing below ground around the root area, provided the twin factors of a gritty soil and a sloping site combine to provide adequate drainage. As the plant ages, however, the rootstock becomes more pithy and absorbent, leaving the roots accessible to frost, and if temperatures then drop low enough to freeze the soil at a depth of 7–10cm (3–4in), the plant is liable to be killed unless mulching has been carried out. The worst danger occurs when heavy rain is followed by severe frost. Air temperatures as low as −18°C (0°F) will be tolerated with impunity, though the plant will to all effects have adopted the habit of a herbaceous perennial.

In temperate zones *E. crista-galli* is sometimes established under glass in a tub and carried outdoors for the summer, in which case annual cutting back is essential to maintain a manageable shape, and this heavy pruning may conveniently be carried out as soon as the leaves have fallen. If heating is available, specimens grown in this manner can with advantage be started into growth very early in the spring, so that they are well advanced before the final frost, and the flowering season will thus be greatly extended. Coral trees should, if possible, be started in an ambient temperature of around 21°C (70°F) until the new growth is fairly forward, when the heat may be reduced to about 10°C (50°F), lessening the potential shock of being carried into the open air after the danger of frost has passed. Without preheating in this way, the coral tree is often loth to flower during particularly cool summer seasons, even if given the shelter of the sunniest wall. After flowering has finished and the annual shoots are ready to be cut back, the tubs should be returned to the greenhouse and the plants kept cool and dry until the spring.

Several shrub and tree erythrinas inhabit the warmer regions of the world, and the vivid red of their flowers is apt to strike the eye in many and varied habitats: the ravines of Nepal; the wooded hill slopes of the Canary Islands; the bush-veld of southern Africa. One of these is the African *E. lysistemon*, which makes a splash of bright scarlet characteristic of the dry winters in many hilly frost-free areas, for flowering takes place in the winter and early spring, the 5cm (2in) curved tubular flowers opening in clusters on the branch tips, or singly here and there on the bare black twigs before the new leaves appear, and a wet summer followed by a reliably dry winter are factors essential for success. To simulate this type of climate when *E. lysistemon* and other woody-stemmed erythrinas are cultivated as greenhouse shrubs in temperate countries, water should be withheld by degrees from early autumn onwards, allowing the wood to harden and the leaves to fall naturally, keeping the greenhouse atmosphere cool and fairly dry throughout the winter. High over-winter temperatures are quite unnecessary, but frost must be excluded or flowering will suffer. The flower buds normally start to open as the temperature reaches 16°C (60°F) or thereabouts. In the earliest spring, container-grown woody erythrinas may be repotted, and those established in a greenhouse border should be top dressed with good loamy soil. During the summer months, a moist atmosphere is essential, and watering should be copious.

In their native lands, the tree erythrinas can be propagated by the simple expedient of taking large branch cuttings or truncheons and thrusting these into sandy or peaty soil. The herbaceous or shrubby types such as *E. crista-galli* may be grown from cuttings, using young sideshoots pulled off with a heel in the spring, as soon as they reach a length of about 7cm (3in). These root fairly readily in a compost consisting of 2 parts sharp sand and 1 part sphagnum peat, with the container set over bottom heat of 24°C (75°F), covering the frame with very light-gauge polythene which is allowed to rest on the young leaves and trap escaping moisture.

Gardenia
Cape jasmine
Gardenia jasminoides (Rubiaceae)

This well-known evergreen shrub, originally from China and Japan, has two quite distinct forms commonly found in cultivation; the one spreading and rather untidy, the other neatly erect up to about 2m (6ft). Both have the typical dark glossy evergreen foliage, and the fragrant, pure white flowers, each 5–8cm (2–3in) across the beautifully symmetrical waxy petals. The flowers on both forms appear somewhat spasmodically throughout the summer and occasionally in the winter too, and both sometimes give rise to a double-flowered strain known as *Gardenia jasminoides plena*. There is also a form with variegated leaves, and this adopts a habit of its own and makes a fairly compact bush, usually under 1.5m (5ft) in height.

In countries warm enough to permit the gardenia to be planted unprotected in the open, as a bush it is characteristically so slow in growing as rarely to need pruning or cutting back; but against this it begins to flower when only around 30–40cm (15in) high, and by the time it has attained 60cm (2ft) or so, is usually flowering regularly and profusely. Old plants flower less satisfactorily than young ones, and when grown in the greenhouse for cut-flower production it pays to regard gardenias as temporary residents only, discarding them after three or four years, and rearing a succession of young plants to provide periodical replacements. Gardenia flowers tend to appear a few at a time on the bush, rather than as a concerted display – a habit which qualifies the species more as a provider of cut flowers for the house than as a display shrub for the conservatory or patio.

Although wild gardenias frequently grow in alkaline soils, G.

jasminoides under cultivation flowers best if given a moderately fertile soil that is slightly acid at pH 6. Like a camellia, it thrives in partial shade and enjoys the generous use of garden compost or leaf mould as a mulch. Traditionally, also like the camellia, it is often given the cold leaves and slops from the teapot to make a fine acid mulch, which is doubtless very beneficial. For greenhouse use, an excellent compost could be made using 3 parts fibrous loam, 3 parts well-rotted manure, 2 parts sedge peat, and 1 part sharp sand. For potting on at all stages, a roughly textured, lumpy compost gives the best results.

Gardenia jasminoides

In order to thrive in their natural state, gardenias need an annual rainfall of at least 60cm (25in), and under garden conditions water should be given fairly copiously whilst the plants are still small. For optimum results under glass, the growing compost should never be allowed to dry out, and the foliage should preferably be syringed each morning and evening in the summer, except while the bushes are actually in flower. When grown outdoors in cool regions they will tolerate fairly sharp frosts on occasion, but under these circumstances the leaves often turn yellow, as though unable during cold weather to assimilate trace elements from the soil. This can be countered with sprays of foliar feed containing trace elements, but growth is much more satisfactory in countries where the winters are always mild. A moist, warm, island climate such as that of Madeira produces perfect gardenias. Conditions in the warmer states of the USA, in South

178

Africa and Australia are often either too dry and hot, or too frosty during the winter on high ground.

Day length seems to make little or no difference to flowering performance, but a cool period of dormancy is beneficial; for greenhouse cultivation, a minimum winter temperature of 13°C (55°F) would be ideal, followed by a summer minimum of 16°C (60°F). The atmosphere should be kept moist in the summer by damping the floor and stages; both air and foliage should be kept as dry as possible during the winter, but watering of the roots must be continued in moderation throughout the plant's resting period. Mature gardenia plants may be started into growth towards the end of winter by raising the air temperature to 16°C (60°F) and, in the case of container-grown plants, setting their containers over bottom heat of 24°C (75°F), increasing the supply of water to the roots and syringing the foliage daily. This treatment will induce free flowering in the spring and early summer. Young, newly potted plants should be given steady bottom heat of around 18°C (65°F) until they are established.

Vegetative propagation can quite readily be achieved using sideshoot cuttings, and this can be repeated at regular intervals throughout the year so as to produce a long succession of flowering plants. Cuttings will root equally well during any month, and a few taken every few weeks will ensure a continuous supply of flowers for most of the year. Midwinter is the best season to take them in the first instance, as the resultant plants will produce enough mature foliage during their long growing period to be able to flower well the following winter and spring. Young sideshoots about 7cm (3in) long should be pulled off with a heel and set singly in small pots, using a compost consisting of 2 parts sphagnum peat to 1 part sharp sand, and incorporating a little superphosphate of lime. The pots should be plunged over bottom heat of 24°C (75°F), keeping them in a close atmosphere until rooting is well advanced, and potting them on – in the case of winter cuttings – in mid-spring.

Hibiscus
Rose of China; shoe flower
Hibiscus rosa-sinensis
also *H. waimeae; H. sinosyriacus* and others   (Malvaceae)

Although not difficult to grow in warm regions, the Far Eastern *Hibiscus rosa-sinensis* is regarded as an exotic prize in most countries of

179

the world. The species, as it is known to gardeners, could more accurately be described as a thoroughly mixed collection of highly polyploidal hybrids which have been cultivated for centuries; many hundreds of named cultivars exist and new ones are being produced annually in countries such as Australia and the warmer states of the USA, particularly California and Florida – the American Hibiscus Society with headquarters in Florida has been formed to cater for the needs of enthusiasts. The prominently stamened flowers vary very widely both in form – single and double, frilled and fringed – and in colour, ranging from white through shades of pink, scarlet, orange, yellow, apricot, salmon, claret, brick red and crimson. In cool regions where they cannot be grown out of doors, large collections of hibiscus cultivars are sometimes kept under glass – as at the Longwood Gardens in Pennsylvania, where they share a house with mimosas and exotic ferns adjacent to a magnificent collection of tropical orchids.

As an outdoor plant, *H. rosa-sinensis* does not require high average temperatures to flourish, and regions in which the winter temperature does not drop below 4°C (40°F) will accommodate them. At temperatures lower than this, the shrub loses its normally evergreen, dark, glossy leaves, though unless badly frosted it will grow new foliage in the spring and still flower well the following summer. The protection of a cool greenhouse is essential, however, in districts where the temperature is liable to drop below freezing point. None of the cultivars will survive a frost more severe than – 3°C (27°F), at least without serious damage; the common, single, red-flowered form is still the hardiest variety (it probably represents the true species), and most of the hybridising work done on the genus has been concerned with establishing strains that have stronger roots and increased vegetational vigour, rather than greater frost resistance. In the Los Angeles State and County Arboretum, where much hybridising work has been done, some slightly less tender ornamental cultivars have been raised, among them Myella, William Stewart and Lasca Beauty, but even these have suffered severe damage and in some cases been killed at the critical temperature of – 3°C (27°F).

In different countries with varying climatic conditions, certain of the flower forms and cultivars are more likely to succeed where others fail, though the differences in performance are slight. Yellow-and bronze-flowered varieties seem to prefer semi-shaded conditions without a lowering of temperature, and thus are not suited to cool regions. Those with other flower colours, including the comparatively

180

Hibiscus rosa-sinensis Cooperi, a variegated form of the Chinese hibiscus, is far less vigorous than the type and more suitable for pot culture in a small greenhouse

Hibiscus sinosyriacus, a valuable late summer and autumn flowering shrub, has been recently introduced to the West from China

hardy single red form, give their best results in full sun. In areas where
− 3°C (27°F) is locally not uncommon, such for example as the high
country of the Transvaal in South Africa, *H. rosa-sinensis* is plainly on
the borderline of hardiness, and even the single red-flowered form
needs the shelter of a warm wall to survive. In subtropical regions not
subject to frost, *H. rosa-sinensis* produces a continuous succession of
long-stamened 15cm (6in) wide flowers throughout the summer,
though each individual bloom lasts only a day or two. In the tropics it
is frequently used as a garden hedge or shelter-belt shrub for street
planting, and will even form a small standard tree if encouraged to
develop a single stem. Commonly making a vigorous 3.5m (12ft) bush
with a spread of 3m (10ft) if left untrimmed, heavy pruning in the
spring is often practised in the garden, even when not grown as a
hedge. A light clipping-over might be carried out at any time of the
year, but the more severe cutting back should be restricted to the early
spring before new growth commences. The variegated form Cooperi
is smaller and far less vigorous − an attractive shrub with its green,
red and white foliage − and more suitable both for the modest-sized
garden and for pot culture in a small greenhouse.

Hardier than *H. rosa-sinensis*, but still requiring a long, warm
summer to excel, *H. sinosyriacus* is a valuable late summer- and
autumn-flowering shrub, fairly recently introduced to the West from
the Lu-Shan Botanical Garden in China. The flowers are usually white
or lilac with a darker maroon or crimson centre, lacking the great
range of colour variation seen in *H. rosa-sinensis*, but very beautiful. It
is reasonably frost-hardy, and enjoys a fairly dry climate with a late,
lingering summer season extending into autumn − an 'Indian
summer' of the type found more often in the hinterland of continents
than in very temperate regions under oceanic influences. Nevertheless,
H. sinosyriacus succeeds in sheltered gardens in the British Isles,
particularly when given the protection of a sunny wall. The cultivars
Ruby Glow and Lilac Queen are particularly recommended for
planting in Britain and Ireland − where fine specimens are to be seen
against the wall which surrounds the now dry moat of Birr Castle, in
the centre of the country, and a district where − 11°C (12°F) is quite
normal in the winter, with an annual rainfall around 85cm (34in).
Another hibiscus which shares this favoured site is the rare *H.
paramutabilis*, also from China, which apparently needs similar climatic
conditions, flourishing at least during the sunniest years, when the
flowers appear in great profusion during August and September,

opening light mauve in the morning and changing intriguingly to crimson at night.

Hardiest of the hibiscus species commonly seen in temperate gardens is the well-known *H. syriacus*, the shrubby althea, also known in North America as the rose of Sharon, immune to the coldest winter weather, and flowering well in a continuous succession during late summer and autumn, superb in warm seasons but satisfactory even in the average dull weather of the UK, with very many cultivars, their flower colours ranging from white to scarlet and violet-blue – not so large in individual blooms as either *H. rosa-sinensis* or *H. sinosyriacus*, but beautiful nevertheless.

Another very attractive species is *H. waimeae*, one of the traditional lei flowers from Hawaii, equal to *H. rosa-sinensis* both in its degree of frost-tenderness and in its beauty, though it lacks the latter's range of flower colour. It is a vigorous plant which needs hard regular pruning when grown under glass. In the temperate house at Wisley Gardens in England it has a long display season during the autumn, with its large, white-petalled, prominently red-stamened flowers joining an exotic group of late-flowering shrubs and climbers, including the pink-flowered *Lapageria rosea* and the dark violet *Tibouchina semidecandra*. In the tropics *H. waimeae* can make a 10m (35ft) tree with a slender, grey-barked stem, but even quite young plants flower freely in climates which suit them. On the French Riviera this species and *H. rosa-sinensis*

Tibouchina semidecandra syn. *T. urvilliana*

183

in many of its colour variations feature in several glorious floral displays with orange thunbergias, yellow bignonias, and the china-blue *Plumbago capensis.*

While the hardy hibiscus species such as the Middle Eastern *H. syriacus* enjoy a sandy, gritty soil, the more tender Far Eastern members of the genus, which probably originated in woodland, prefer a fibrous compost fairly rich in humus. Within temperate zones, especially when grown in large containers to stand outside for the summer, *H. rosa-sinensis* can be given a moderately fertilised compost made up of 2 parts turfy loam, 2 parts fairly coarse sphagnum peat and 1 part sharp sand, plus an admixture of ash and small pieces of charred wood from a bonfire. After keeping the plants fairly dry and at a temperature of 4–7°C (40–45°F) over winter, watering in the spring should be accompanied by thorough damping of the surrounds to produce a moist atmosphere, and if possible bottom heat of around 24°C (75°F) should be given as the outdoor temperatures start to rise. Cutting back, if required, can be done at the same time.

H. syriacus is usually grown from fairly hard cuttings about 10cm (4in) long, taken in the autumn and kept closely covered over bottom heat of 18°C (65°F), when they should root within a fortnight or so. In warm regions *H. rosa-sinensis* and the other tender species are also fairly easy to grow, either from hardwood cuttings taken during the winter, or from soft tips or sideshoots taken at any time during the growing season. Seed of *H. rosa-sinensis* germinates readily, but to obtain true cultivars or fancy varieties, vegetative propagation is necessary. Within cool zones subject to frost, cuttings of new growth, about 8cm (3in) long, may be taken in the spring and set in an evenly balanced peat/sand compost over bottom heat of around 24°C (75°F), watered well and covered with scored polythene which must be stretched to admit air as soon as new growth is apparent. When rooting is well advanced, the cuttings should be potted into individual 10cm (4in) pots, using a lightly fertilised version of the container compost used for mature plants.

Lagerstroemia
Crape myrtle; pride of India; Indian lilac
Lagerstroemia indica   (Lythraceae)

A native of China despite its Indian specific epithet and popular names, this beautiful shrub has 4cm (1½in) crinkle-petalled lilac-pink flowers produced during the summer and autumn in huge, upright plume-like sprays. Crape myrtle can assume the proportions of a 9m (30ft) tree in those subtropical areas where it is most at home, but is more frequently to be seen as a 4.5m (15ft) shrub, the branches springing from the base and arching outwards to reach a spread equal to the height. The small, thickly textured dark green leaves sometimes remain on the tree during tropical winters, but elsewhere they are deciduous, and often adopt handsome autumn colours before falling. The flowers appear on the current year's growth and, as the tree is amenable to heavy pruning and trimming, branches may be cut hard back in the winter and early spring, in the manner of *Buddleia davidii*, without loss of flowering potential. Alternatively, it can be trained as a small standard tree and grown almost in the topiary style by clipping a neatly rounded crown which annually becomes covered with flowers. The paper-thin bark is ornamental, too, satin-sheened and mottled in shades of grey, pink and fawn.

Flower colour is variable; naturally occurring varieties with crimson, pink or white flowers often appear, and seedlings do not necessarily come true to their parents' colour. It has often been remarked that the common mauve or lilac type comes first into flower, and is also first to lose its leaves in the autumn. When grown outdoors in temperate climates the vagaries of the weather have some effect on flower quality, and it seems likely that the typical lilac-coloured form will do best in these regions. After flowering, the dried seed husks remain on the bare branches over winter and are quite ornamental.

Magnificent specimens of crape myrtle are to be seen near the Mediterranean. In Israel the species makes a tall deciduous shrub, flowering profusely during the summer and autumn. In that warm and arid climate it is one of the few plants to produce conspicuous autumn coloration of red and orange, and the young leaves are attractive too in the spring when they have a reddish tinge. Further west, the species finds the climate of southern Europe ideal, and it has no objection to a modicum of frost. It can be grown on sunny sites even in fairly cold regions, provided some winter protection is given in the form of a mulch for the roots, and lagging for the lower stems – trees which have been grown as single-stemmed standards are the easiest to protect in this way. The length of the summer season is as important as its temperature and, given winter protection, crape myrtle often proves more reliable outdoors in the eastern coastal states of the USA than in the far more temperate climate of Britain, although some fine specimens exist particularly in the south and west of the UK. Beautiful crape myrtles are to be seen for instance in Cornwall and Devon, and also in the famous woodland gardens at Bodnant in North Wales. In Hillier's Arboretum near Winchester, England, the species grows well in front of a sheltering wall, but the flowering performance here would undoubtedly be much better were the summers longer and hotter.

Outdoors in the New England states crape myrtle makes a vigorous, free-flowering bush for sheltered gardens, and at the University of Pennsylvania Arboretum it has grown into an impressive 9m (30ft) tree and flowers beautifully. It is also grown as a small tree at Dumbarton Oaks in Georgetown, Washington DC, annually putting on a spectacular display after most of the flowering trees and shrubs have finished. Further south, in the historic gardens at Charleston, South Carolina where, though occasional sharp frosts occur, the winters are never severe, it makes a large multi-stemmed tree amid the magnolias, camellias and rhododendrons, competing with the mossy live oaks and huge bald cypresses around the lake.

In the open, crape myrtle is at its most picturesque in the form of a free-branching standard tree, but when grown under glass specimens are best kept clipped to manageable proportions, particularly if established in movable containers. A happy compromise is seen in the enormous cool conservatory at the Longwood Gardens near Philadelphia, where the species is grown as a closely clipped standard to form the centrepiece for a display of tender and half-hardy bedding plants.

Seed is best sown as soon as it ripens in the autumn, using a compost of 2 parts loam, 1 part sphagnum peat and 1 part coarse sand, adding a little superphosphate and a sprinkling of ground limestone. Cover the seed lightly and water very gently, keeping the seedbox shaded and covered with glass or polythene, and set over bottom heat of 24°C (75°F). When germination has taken place the shading should be removed and the seedbox stood as close to the greenhouse glass as possible to ensure maximum light during the winter. In the spring, the seedlings should be potted individually, using a compost of 3 parts turfy loam, 1 part sphagnum peat and 1 part coarse sand to which has been added a fair proportion of bonemeal.

As seedlings do not come true to colour, the best flower colour variations need to be propagated vegetatively. Cuttings of half-mature

186

sideshoots some 5–7cm (2–3in) long with a heel can be taken in the late spring and early summer, at the stage when they are firming at the base, and set singly in small pots containing 2 parts loam, 1 part sphagnum peat and 1 part sand, allowing a little fine sharp sand sprinkled on the surface to drop around the heel to assist rooting. The pots should be watered well and plunged in bottom heat of 27°C (80°F), the frame covered with scored polythene which can be stretched to admit air as rooting takes place. Alternatively, root cuttings can be taken early in the winter, using pieces about 4cm (1½in) long and 1cm (⅜in) diameter. They should be placed horizontally 2–3cm (1in) below the surface in small individual pots half filled with compost as recommended for shoot cuttings, but with the roots set in a layer of pure sharp sand, and the pot topped up with compost. The pots should then be plunged in sand over bottom heat of about 24°C (75°F), which can be reduced to about 18°C (65°F) when new top growth appears.

Leucadendron
Silver tree
Leucadendron argenteum and others   (Proteaceae)

Leucadendron argenteum from South Africa is usually considered one of the world's most beautiful trees; though not by merit of its flowers which are individually inconspicuous. The long, tapering bracts which surround the tightly clustered flower heads for four or five months of the year during the Cape winter and early spring, are quite ornamental; but it is in the foliage where the beauty lies. The leaves are pointedly oval, 10–12cm (4–5in) long, and covered with silky hairs which glisten silver in the sun and give the tree its popular name. These hairs are raised during rainy weather to allow maximum absorption of moisture, but when hot, dry winds blow from inland they lie flat to protect the surface of the leaf from excessive transpiration.

On Table Mountain they are spectacular trees, symmetrical and sentinel-like on still days, but rippling as often as not in the prevailing breezes, their silvery foliage contrasting vividly with the dull grey of the young stems. Introduced specimens of the silver tree are to be found in subtropical countries everywhere, and the species is a favourite garden tree in Australia and parts of New Zealand. Following its introduction to California it has found the local climate and soil so much to its liking that it has become naturalised in many places and grows there freely, some specimens exceeding even the native trees at the Cape of Good Hope in size, reaching 7.5m (25ft) or occasionally 9m (30ft), though seldom spreading more than 4.5m (15ft) with their high-angled branches. In California, as at the Cape, it is a short-lived tree, and seldom survives longer than twenty years or so. Even in their native habitat, apparently youthful silver trees will sometimes die for no apparent reason; deaths have often been attributed to drought at the end of the growing season (at the Cape the summers are dry), but many unexplained deaths of garden trees in the past could have been explained simply in terms of old age. Not a very rapidly grown tree, it may take four or five years to reach 2m (6ft) in height, and the size of a well-grown specimen will give a reasonably accurate indication of its age.

Leucadendron plumosum

In the wild, the silver tree prefers to grow in comparative isolation rather than in groups, favouring sites with a deep, gritty soil, chiefly on steep mountain slopes where the drainage is thorough and the humidity, especially during the winter, is high; but it is fairly resistant to drought and heat, and, at the other extreme, will tolerate a few degrees of frost. Away from its native habitat, gravelly soils or coarse sandy loams produce the best results; it will not succeed in heavy clay, or on sites where the drainage is poor. In climatic zones with mainly summer rainfall, young trees are liable to need watering during autumn and winter, at least for the first two or three years after planting out. Moist, mild breezes help to supply moisture via the foliage, and the hairy leaves are well adapted to cope with hot desert winds; cold winds on the other hand, whether damp or dry, often prove fatal.

Many other leucadendron species originate in southern Africa, all of them smaller and less showy than *L. argenteum* – though the shrubby *L. plumosum* and *L. stokoei* are very attractive and sometimes grown as greenhouse plants. None of them will survive a typical winter outdoors in mainland Britain, though both *L. plumosum* and the silver tree itself have been planted in the Scilly Isles where, in the gardens of Tresco Abbey, *L. argenteum* attains some 6m (20ft) and flourishes amid a collection of trees native to the southern hemisphere: casuarina, *Puya chilensis*, aloes and proteas, tree ferns and *Agathis australis*. There, the mild Atlantic breezes and the almost complete absence of frost create conditions not dissimilar from those found near Table Bay.

Seed is the normal method of propagating *L. argenteum*, and it should be noted that the trees are either all-male or all-female. In South Africa, sowing is usually carried out in the autumn, using a light sandy soil to which humus has been added. The seed is covered no deeper than 1cm (¼–½in), and it is not thought necessary to shade the beds or boxes from the sun. In hotter climates a modicum of shade might be given at least until after germination, which occurs in three or four weeks. The seedlings are potted up into a sharply gritty but fairly humus-rich compost as soon as they have developed their first true leaves, and after pinching out the tips to encourage bushiness are planted in their final positions when about one year old, or 30cm (12in) high. Older plants do not transplant well.

In temperate zones, imported seed is normally sown under glass in the very earliest spring, using bottom heat of about 21°C (70°F). Seedlings have proved highly susceptible to damping off in damp climates such as that of the UK, and to avoid this infection, dust-free sieved clinker has been used to top the seedboxes or pots. Containers should be crocked to a good depth to ensure perfect drainage, and the compost, which might consist of 2 parts sharp sand to 1 part sphagnum peat, is topped with 1cm (¼–½in) layer of clean, pinhead-size clinker. The seed is sown on this and covered with a further shallow layer of clinker. The propagating case should be left open and well ventilated; after germination shade should be given on sunny days, but when the seedlings are about 2–3cm (1in) high they must

189

gradually be exposed to full sunlight. A suitable potting compost consists of 2 parts turfy loam and 1 part sphagnum peat to 1 part coarse sand, with clinker again used as a surface layer, taking care not to bury the root collar. As a further precaution against damping off, it is advisable to add Cheshunt compound to the water (11 parts ammonium carbonate to 2 parts copper sulphate) finely powdered and diluted at the rate of 3g/litre (½oz/gallon) of water, for use from the pre-germination stage until potting on.

Mimosa

Sensitive plant; humble plant
Mimosa pudica
also Acacia; wattle
Acacia dealbata; A. baileyana and others   (Leguminosae)

The group of plants usually known as mimosa, and sold under that name by florists, comprises not the true mimosas but the acacias – often *Acacia dealbata*, the silver wattle. To the genus *Mimosa* belongs the sensitive plant, *M. pudica* – although, to be true to the Latin, *M. pudica* should correctly be known as the humble plant, the name sensitive plant belonging to the less sensitive *M. sensitiva*. Both are natives of tropical America, and have become naturalised here and there in tropical regions throughout the world. They are very tender, and need greenhouse protection wherever temperatures are liable to drop in the winter below 18°C (65°F), and in the summer below 21°C (70°F).

M. pudica is a short-lived perennial usually grown for convenience as an annual, and makes a prickly but not very robust little plant, usually 30–40cm (12–18in) tall, with finely compound leaves and little pinkish purple flower-globes appearing in the summer. A light touch on the tip of a leaf is enough to cause all the leaflets to droop and fold closely together in pairs, a continuous movement spreading like a ripple along the leaf and continuing rapidly from stem to stem until the whole plant has wilted. Under greenhouse cultivation it needs maximum light, and does best in a comparatively small pot with a lightly fertilised compost, which might consist of 4 parts loam to 2 parts sphagnum peat, and 1 part sharp sand.

The sensitive plant can be propagated from cuttings taken at any time of the year and set in a warm case, but it is usually grown from seed. This germinates more quickly than that of the acacias, and no

Mimosa pudica, the sensitive plant, is usually grown as an annual, cultivated not for its beauty but for its curiosity value

pre-treatment is required: it should be sown in a compost consisting of 1 part loam, 1 part sharp sand and 1 part sphagnum peat, and set over bottom heat of 24°C (75°F).

The true acacias are mostly far hardier than mimosa proper, but even the robust Australian silver wattle *Acacia dealbata* will not grow entirely unprotected in places where the winter temperatures are liable to drop much below freezing. Specimens will survive an occasional low of −6°C (21°F), but when grown in temperate zones such as the UK are safer given the protection of a sunny niche in which they can be sheltered from cold winds and the effects of late frosts. In regions

Acacia dealbata

such as these, though *A. dealbata* will grow outdoors as a wall shrub, say in the south-west of England, frost often spoils the flowers and sometimes kills the branches back to the main stem, robbing the tree of its symmetry; the best specimens therefore are usually to be seen under glass, either in large pots or planted permanently in a greenhouse border. In subtropical countries, it might come as a surprise to learn that the species needs cosseting elsewhere, for it can prove difficult to eradicate once firmly established; it suckers strongly and can reach a height of 9–10m (30ft) or more, growing vigorously even in very poor soil. When planted outdoors in areas subject to frost it is much better behaved, and can well be grown as a handsome wall plant in colder districts where its survival over winter is largely a matter of chance, replacing specimens as necessary by propagating them regularly and keeping spare plants under glass.

Acacia baileyana, Bailey's or Cootamundra wattle from New South Wales, flowers in the winter and spring, and is much better behaved than the silver wattle, never growing too rampantly in tropical or subtropical gardens. It makes a neat and highly ornamental little tree, with feathery foliage of a waxy, silvery blue, and great fluffy clusters of lemon-yellow flowers. It is one of the most commonly grown mimosas on the French Riviera, and a popular garden plant in warm countries the world over. It will tolerate some frost, and can be grown outdoors as a wall shrub in sheltered gardens in the milder parts of Britain – but even in these districts it is safer kept under glass where frost can be excluded. A little artificial heat, in any case, is highly beneficial for the flowers which open during late winter, and a minimum temperature of 10–12°C (50–55°F) results in the best flowering performance. In climates which really suit Bailey's wattle, it can attain a height of some 7.5m (25ft), with a spread across the branches of some 6m (20ft).

In the case of small trees like the acacias, sacking can very effectively be used to protect the stems overwinter, lagging it loosely around both the main stem and the crotch formed by the largest branches, as far as possible protecting a solid base on which a new crown can be built should it become necessary. Very small trees or shrubs can sometimes be completely enclosed in a large sack supported on a tripod, or twiggy sticks can be arranged to form a tent which is thatched with

One of the most commonly grown mimosas, *Acacia baileyana*, Bailey's or Cootamundra wattle from New South Wales, makes an ornamental little tree

straw or bracken. Sacks used for the purpose should always have been washed clean, and be free from lime or chemicals.

In America, the New England states experience weather far harder than that of southern English winters, yet very often plants apparently on the borderline of hardiness are able to flourish there. The longer and more continental summer season allows the current year's wood to ripen fully before autumn, so that it remains safely dormant before breaking cleanly into growth when the temperature rises in the spring – the flower buds opening later then they are apt to do in more temperate regions, with less consequent risk of damage from late frosts. In those parts of the British Isles under the benign influence of the warm Gulf Stream some interesting specimens of acacia are to be found: on the south coast of Ireland, clinging to the walls of an ancient castle, *A. dealbata* reaches some 7.5m (25ft) in the company of a winter-flowering *Buddleia auriculata*, a white wistaria, and the sky-blue *Ceanothus cyaneus*; and in a nearby sunny but sheltered woodland garden, the same species has reached an impressive 15m (50ft), and flowers regularly each spring.

The Australian blackwood acacia, *A. melanoxylon*, is a handsome tree with heavily scented though not very beautiful flowers. Like *A. dealbata* it suckers profusely in tropical and subtropical climates, and in some parts of the world has become a menace to agriculture, so difficult is it to eradicate once established. In the temperate zones it has rarely been thought suitable for greenhouse culture, but it is moderately hardy outdoors, growing freely in Cornwall, England, where it can attain 21m (70ft) or more. Though somewhat susceptible as a young sapling, once it has passed 3m (10ft) or so and made a firm trunk it will withstand severe frosts without damage. There are several good specimens on the windswept coast of County Kerry in the extreme south-west of Ireland, where 21m (70ft) blackwoods stand over thickets of their own self-sown seedlings. The flower buds have already been formed by midwinter, and cold winds are liable to cut them back, but in mild seasons the creamy yellow blackwood blossom often perfumes the surrounding air in the early spring.

A species of equal hardiness to the blackwood is *A. longifolia*, the Sydney golden wattle, highly ornamental in the spring with its bright yellow flowers – a tree which could be planted far more often in comparable sites. *A. riceana* from Tasmania bears pale creamy yellow flowers in great profusion from early to late spring, one of the most graceful of acacias with its weeping clusters of blossom. Reaching

194

Buddleia auriculata

Acacia riceana

195

around 9m (30ft), it can be magnificent when trained as a wall plant, and this, too, is moderately hardy, tolerating temperatures as low as $-4°C$ (25°F).

The acacias tend to develop deep taproots as a natural precaution against drought, and for this reason should be transplanted while still very small. Typical pioneer species in the wild, particularly those species from Australia and Tasmania, they are often the first trees to appear on the blackened ground following a forest fire, for the hard-coated seeds germinate freely after they have been scorched in the flames. This process may be simulated to speed up germination: the seeds should be spread thinly on a sieve which is then passed over a flame until some of them explode audibly; at this stage they should be plunged into lukewarm water for a few minutes, then sown without delay in a compost consisting of 2 parts fine loam, 1 part sharp sand and 1 part sphagnum peat, containing superphosphate but no free lime, and covered lightly with a sprinkling of pure sand. The seedbox should be stood over bottom heat of 21°C (70°F) and covered with paper until germination takes place, and the seedlings left undisturbed until rooting is fairly well advanced, when they can be potted up into tall 10cm (4in) pots or tubes, using a moderately fertilised sandy compost. As a good alternative method of preparing acacia seed for germination, the batch may be placed without scorching into water that has been brought barely to boiling point (100°C or 212°F), and left to soak overnight in the cooling water. The following day they should be rinsed thoroughly with clean water and dried in the shade before sowing as previously described. In both cases, sowing should take place as soon as ripe seed has been obtained, irrespective of season.

Acacia plants grown from seed tend to give better results in the garden than those raised from cuttings, but if no seed is available the vegetative method can be used, taking half-ripened 8–10cm (3–4in) sideshoots with a heel, soon after midsummer. The cuttings are best set singly in small pots plunged into sand over bottom heat around 16°C (60°F), and an ideal compost might consist of 1 part loam, 1 part sphagnum peat and 1 part sharp sand. They should be kept close until rooting is well advanced, before moving them to an open green-house to grow on under cool conditions.

A. melanoxylon is usually grown from root cuttings: pieces of root are dug up during the winter and washed free of soil, cut into 2–3cm (1in) lengths and set singly in small pots. A compost of 2 parts loam, 1 part sphagnum peat and 1 part sand is suitable, the root cuttings placed horizontally and covered shallowly. The pots should be watered well and plunged in sand over bottom heat of 21°C (70°F), keeping the frame closed and well shaded until new rooting is apparent. It is better to leave root cuttings undisturbed until they are quite well developed, by which time spring will have arrived, and the pots can be transferred to an open greenhouse.

Oleander
Rose bay; poison bush; Ceylon rose
Nerium oleander   (Apocynaceae)

By the shores of the Mediterranean and in its native southern Europe, this sun-loving evergreen rivals the camellia of more northerly gardens, bearing large clusters of 4cm (1½in) waxy-petalled flowers from early summer until the autumn, in shades of pink, red, white or sometimes yellow, amongst leathery greyish-green lance-shaped leaves. Erectly branched and round-headed whilst young, oleander spreads bushily in maturity until eventually it needs a great deal of room, attaining some 4–5m (12–16ft) in height and as much across. It is grown very extensively throughout the world in tropical and subtropical countries, and is valued as much for its useful evergreen foliage as for its gorgeous flowers. There are variegated forms which are doubly attractive in the garden. The pink- and white-flowered types especially are familiar shrubs within its native range, and in Greece and southern Turkey there are many wild wooded valleys where it flourishes and seeds itself freely along the stream banks and amongst the trees. A favourite summer flowerer in Israel too, it grows well in poor soil, and tolerates both drought and a modicum of frost. These characteristics make it a choice garden shrub for Australia and South Africa, where the climate varies greatly from district to district, and a favourite choice particularly for coastal planting in such countries, when the flowers open in the early spring and continue in succession throughout summer and well into autumn.

The sap of the oleander is extremely poisonous, and both cattle and horses have died after eating the leaves. The flowers are reputedly poisonous, too, and humans are said to become ill after eating them. The somewhat unlovely name of 'poison bush' used in many countries acts as a deterrent both to children and to adults unfamiliar with these properties.

In cool climates oleanders need the maximum of sun and air during spring and early summer if they are to flower well later in the summer, as flowers are produced only on mature, well-ripened shoots, and cosseted greenhouse plants which are too soft of growth sometimes fail to bloom well for this reason. Under glass in temperate zones it is usual for the flowers to open around midsummer and last for two or three months. After flowering has finished the water supply should be withheld for a month or so until mid-autumn, and any necessary trimming or pruning should be done then. In the subtropics where growth is very rapid and flowers are freely produced, trimming may also be carried out in the early spring if required. In mid-autumn, watering should be restarted and the foliage syringed in order to induce a final flush of growth before winter sets in, as this will assist the ripening of shoots to flower the following year. During the winter, although a resting period, enough water should be supplied to maintain the compost in a moist condition, bearing in mind that, as a Mediterranean native, oleander in nature experiences a winter rainfall. During the winter, an average temperature of 10–13°C (50–55°F), with a minimum of 7°C (45°F), will ensure the healthiest plants. As high temperatures are not needed for flowering, artificial heat is seldom required during the spring or summer.

Oleander is one of the most suitable flowering evergreens for tub culture within cool temperate zones, and greenhouse plants often give of their best when carried outdoors to spend the summer on an airy, sunny patio, returning to the conservatory before the first frost is due. They can make very attractive small standard trees in this situation. Repotting, when required, should be carried out in the early spring, using a compost of 2 parts sandy loam to 1 part decayed manure. An occasional liquid feed containing trace elements should be applied during the growing season.

Oleander roots easily from cuttings. Half-mature sideshoots with or without a heel may be taken at any time during the growing season, but preferably after they have become ripe and firm at the base, and between 8–10cm (3–4in) long. For convenience they can be set into individual small clay or paper pots, in a compost consisting of 2 parts sphagnum peat, 1 part loam and 1 part sharp sand, surfaced with pure sand which will trickle down to the base to assist rooting. They should be watered in firmly and the pots set under a thin polythene sheet, over bottom heat of about 24°C (75°F). In subtropical climates, cuttings are often persuaded to root by standing them in water for a few weeks, and this method can be equally effective in temperate

A double flowered form of the rose bay, *Nerium oleander*, makes an attractive tub plant for conservatory or patio

countries: in the case of oleander, long surplus shoots can be used after trimming them off in the autumn; left in a jar of water on the windowsill over winter they will usually have produced roots by the spring, when they can be potted up in a compost of the type described for conventional sideshoot cuttings, and given a little bottom heat until they are well developed.

Protea
Sugar bush
Protea cynaroides and others   (Proteaceae)

Proteas are to be found throughout Africa as far north as the hills of
Ethiopia, but most of these evergreen shrubs or small trees are native
to the south-west Cape, where they are well known collectively as the
national flower of South Africa. The globular flowers are made
conspicuous by the broad, cup-shaped arrangement of bracts which
surrounds them, in a range of colours from white through shades of
pink and red to purple and black, often silvered and sheened with a
covering of silky hair. The best species for garden use come from the
Cape, where the well-rounded, spreading bushes with their smooth,
olive-green, leathery leaves, displaying a splash of colour here and
there, form a familiar part of the landscape. The flowers are rich in
nectar, and *Protea mellifera* – the true sugar bush known in Afrikaans
as 'suikerbos' – was used by early European settlers there in the
manufacture of sugar. Not one of the most beautiful of the genus, *P.
mellifera* is nevertheless attractive, with colours ranging from off-white
through shades of pale yellow to a deep crimson pink, tinting the
bracts which flank the long, narrow, waxy flower heads, on show in
the Cape from mid-autumn to the spring. A 2.5m (8ft) evergreen
bush, it usually thrives in soil often little better than pure sand.

Flowers of other protea species range in size from the tiny *P.
lacticolor* to the giant, artichoke-like *P. cynaroides*, up to 27cm (11in)
across, and known as the giant or king protea, bright pink with a
silvery sheen, on display in the Cape autumn from January to June,
and sometimes again in September. One of the easiest to grow is *P.
compacta*, with 10cm (4in) pink heads; *P. neriifolia* has long, pink bracts
tipped with silky black hairs; *P. barbigera*, the large woolly-headed
protea or giant woolly beard, blooms through the winter on high
ground and will withstand fairly intense cold, making a spreading
bush 1.5m (5ft) high and some 2.5m (8ft) across, with 13cm (5in)
heads of yellow, brown and pink woolly-haired bracts. A marvellous
collection of proteas is to be seen in the South African National
Botanic Gardens at Kirstenbosch.

Even in their native habitat, well-grown proteas occasionally die for
no very obvious reason – sometimes, probably on account of old age,
but more frequently sudden deaths are said to result from drought at
the end of a long, dry summer. Waterlogging is certainly fatal for

200

Protea cynaroides

them, and when cultivating proteas it is essential to provide a well-drained site – perhaps in the form of a raised bed built in an open, sunny position, with a light soil that contains a certain amount of humus in its top layers, but with a deep gravelly subsoil beneath. Accustomed in nature to a moderate winter rainfall, proteas must be watered regularly during the winter when grown under glass, and kept barely moist in the summer. Although some protea species from the tropics grow naturally in areas of summer rainfall, the habitat in such cases is almost invariably to be found on the dip slope of high escarpments, or the leeward flank of mountain ranges, sites which are subject to frequent mist and drizzle throughout the winter. In the mountains of the Western Cape Province, where the native species normally grow on poor, gravelly soil, rainfall tends to be copious throughout the year, and the usual flowering season lasts from late autumn through the winter until mid-spring. Most of the proteas will tolerate fairly severe frost when fully grown, provided the soil does

201

not become frozen around the roots; plants of five years old or more will withstand temperatures as low as $-6°C$ (21°F) without damage, but even a light frost can prove fatal while the crown branches are still soft and sappy during their first two or three years of life.

P. cynaroides, the most popular garden species, can be grown as a cool greenhouse plant where climatic conditions are too extreme for outdoor planting. Areas in which the winter minimum temperature averages above 5°C (41°F) and is unlikely to drop below the crucial $-6°C$ (21°F), and in which plant growth normally occurs over at least eight months of the year, will support the king protea outdoors provided the soil is acid and well drained; specimens have survived several seasons outdoors in the north-west Highlands of Scotland, where the climate is influenced by the warm Gulf Stream. In the much warmer coastal regions of California, proteas are grown extensively and seem well adapted to the climate, provided they are watered during the summer. There *P. cynaroides* makes a handsome 3m (10ft) bush with sprawling branches, and clumps of the species are sometimes planted closely to form a dense thicket, well covered with silvery-pink flower heads during the Californian late summer and autumn.

Proteas resent root disturbance once they have become established, and cultivation near the plants is undesirable. When the roots are left undisturbed growth is fairly rapid, the larger species reaching perhaps 2m (6ft) in four years, and they commence flowering well from their fourth or fifth year. A light, sandy, acid loam with the addition of leaf mould is ideal, but proteas are sparse feeders and resent the use of manure, unless it is very old and well decayed. In any case, a well-drained sunny slope provides the ideal site.

Proteas can be propagated from cuttings: in the case of greenhouse plants within northern temperate zones, almost fully ripened sideshoots should be taken in midsummer, either at a node or with a heel of older wood, whichever is convenient. In terms of drainage, clay pots are easier to manage than plastic ones, and they need to be well crocked, placing a mixture of peat and coarse sand directly over the crocks, but with pure sharp sand forming the medium into which the cuttings are inserted around the rim of the pot. A convenient pot diameter is 15cm (6in), and care should be taken to ensure that the sand has a neutral or slightly acid pH reaction. The pots should be given bottom heat of 18°C (65°F) and the frame kept closed, or covered with thin scored polythene which is stretched to admit air at the first signs of new growth, and the cuttings potted individually as soon as rooting is well advanced, using a lightly fertilised mixture of peat and gritty sand in equal parts.

Plants grown from cuttings, however, are often said to have inferior root systems, a fact which becomes apparent later in life, and if seed is available it should be used in preference. Protea 'nuts' take a full year to ripen, and if seed is to be collected, the dead flower heads should be left on the bushes until the following autumn, when they are harvested and the seed rubbed out from the dried bracts. In the normal course of events, and if seed is not required for propagation, the dead heads are best removed as they fade to conserve plant energy and encourage vegetative growth.

In California seed sowing takes place in the autumn as soon as harvesting is complete, in boxes or beds made up in a cool, airy greenhouse, using a compost consisting of equal parts sand and well-screened sphagnum peat. Germination follows after one or two months, and the seedlings are potted into 7cm (2½in) pots as soon as they are large enough to handle, and immediately stood outside under a lath shelter. To prevent damage to the lateral roots proteas are never allowed to become crowded in their pots, but are moved into larger sizes as required until ready for final planting out in the spring. In South Africa, sowing is usually carried out in the autumn in open seedbeds, or sometimes small cultivated patches in which the plants are to grow permanently, and germination time is rather variable. For open seedbeds a light, acid soil with completely free drainage is essential, and shallow sowing gives the best results, setting the nuts no deeper than their own diameter. It is important to maintain constantly moist conditions for germination, and it is thought that the seeds contain a germination inhibitor which must first be washed out by copious watering. As soon as the initial pair of true leaves appear, the seedlings are transplanted very carefully so as not to damage the superficial side-roots.

Under glass within the cooler temperate zones, protea seed is sometimes sown into a 1cm (⅜in) thick layer of pure clinker, sieved cleanly to about pinhead size, and covered lightly with the same material. Below this surface layer in the seedbox a compost of 2 parts coarse sand to 1 part sphagnum peat is set over a deep layer of crocks to ensure perfect drainage. Using bottom heat of 18°C (65°F), germination usually takes place after two months, and both seeds and seedlings are watered regularly with a solution of Cheshunt compound (2 parts by weight finely ground copper sulphate to 11 parts fresh ammonium carbonate), allowing the chemicals to interact dry for twenty-four hours before dissolving the mixture in water at the rate of 3g/litre (½oz/gallon). These precautions stem from the need for copious water at the pre-germination stage, combined with the great susceptibility of protea seedlings to damping off under greenhouse conditions, and in cool, humid climates. The clinker serves to keep the root collar comparatively dry, the peat/sand compost retaining water at the roots while allowing free drainage. Full sun and air should be admitted once the seedlings are 4cm (1½in) high, and potting up should be completed before they have exceeded a height of 5cm (2in), using a light, sandy compost.

Telopea
Waratah
Telopea speciosissima; T. oreades; T. truncata and others   (Proteaceae)

Telopea speciosissima is from New South Wales, where it is recognised as the national flower; an erect evergreen shrub with leathery leaves 20–25cm (9in) long, the large deep red flowers are crowded in terminal clusters, like the proteas to which they are related, the 8cm (3in) curled bracts forming a globose flower head which may measure some 10–15cm (4–6in) across. The generic name telopea refers to the distance from which these spectacular flowers may be seen, glowing coral red or crimson – though white and pink varieties are occasionally to be found – during the spring and early summer in the Australian bush. Usually some 2.5m (8ft) high, the waratah can attain a height of 3.5m (12ft) with a spread up to 2m (7ft).

In nature it inhabits only very restricted areas, and the other members of the genus are also of limited distribution. *T. speciosissima* itself occurs in a relatively small region of sandstone country near the east coast of New South Wales, where it crowds into small isolated clumps ranging from 0.5–4 hectares (1–10 acres) in extent, each waratah grove separated from its neighbours sometimes by several miles. *T. mongaensis* is even more local in distribution, and restricted to a small and rather cold area of hilly country in New South Wales about 300km (200 miles) inland, where this fairly hardy tree also grows strangely isolated in self-contained groups; and in another hilly region some 800km (500 miles) to the south the hardier *T. oreades*, which can make a tree approaching 15m (50ft) in height, also shares the characteristic habit of growing in small, isolated clumps, scattered in this case over a far wider area. By contrast, *T. truncata* from Tasmania is evenly and widely distributed over a wet, mountainous area and, like other Tasmanian trees, is moderately hardy.

Of these species, the crimson-flowered *T. oreades* will survive in those temperate regions in which the winter temperature averages 6°C (42°F) or more, provided it does not encounter cold spells below

Telopea speciosissima

205

$-8°C$ (18°F); it succeeds outdoors in Cornwall, England. *T. truncata* is more tolerant of cold spells, especially when planted amongst evergreens in an acid, peaty, moist yet well-drained soil − a typical woodland garden habitat in southern Britain − where it makes a large shrub or sometimes a small tree, with stout downy shoots carrying crimson-bracted flower heads in early summer; but it appreciates a site which allows sunshine to strike the thick evergreen foliage, whilst the roots remain coolly shaded and sheltered beneath a mulch of leaf mould or peat. *T. truncata* survives as far north as Kilbride in Northumbria, where it has been seen on a housewall facing the rising sun − but here it picks up some heat from a chimney behind the brickwork, and is saved from the biting east winds by being well covered with sacking during the winter. In the even more northerly, but milder climate of the Western Scottish Isles, the Tasmanian waratah grows without wall protection in a woodland setting, and has reached an impressive 6m (20ft) or more in company with other exotics from the southern hemisphere, such as embothriums and eucalypts; and in other parts of the British Isles where frost is not too severe − as near the coasts of Ireland − specimens at least 4.5m (15ft) tall may occasionally be found, their dark green leaves adding to some woodland garden scene, their crimson whorls of bracts extending the rhododendrons' flowering season.

In the more sheltered gardens within these favoured places, though sometimes difficult to establish − partially on account of a strong dislike of root disturbance − *T. speciosissima* itself occasionally succeeds if allocated a sunny site, particularly one on the edge of those shrubby tangles of vegetation which resemble the waratah groves it favours in Australia. Provided the soil is rich, moist and free from lime, specimens tend to excel in sheltered but sunny gardens which have become wild and neglected. In such a situation, in balmy Cornwall, the flowers are open sometimes by mid May and last well into the summer.

The telopeas set seed freely in the wild, but the pods tend to be produced only biennially and, when they appear, a season's flowering is missed the following year. For this reason, if seed is not required for propagation, spent flower heads should be picked off before they can set. Fertile seed often seems unwilling to germinate, and remains viable only for about three months. Australian gardeners believe there to be a microscopic symbiotic fungus or mycorrhiza associated with the roots of mature, established telopeas, which would account both for their natural restriction to isolated groups, and for their poor germination in the nursery. They will produce roots fairly readily from both stem and leaf cuttings, but the resultant plants often prove unwilling to grow away, as though some element were lacking in their diet. There may be some advantage in including a little soil taken from among the roots of established specimens when making up the sowing or rooting compost, so that any minute fungi will be introduced.

In Australia, waratah plants are often grown from root cuttings, setting small pieces of lateral root in individual paper pots with a compost of sand, leaf mould and sphagnum peat, over bottom heat of around 18°C (65°F). Established plants may also be layered during the second half of summer or early in the autumn, using 15cm (6in) pots plunged below soil level at convenient points around the parent plant, filled with a compost of 2 parts sphagnum peat, 2 parts coarse sand and 1 part well-rotted manure, and embodying a little of the surrounding soil. Shoots of the current season's growth should be selected, giving them a slight twist to break the bark at the point to be rooted – preferably about 15cm (6in) from the tip. Older shoots with more brittle wood can be used, but instead of twisting, a tongue should be cut about 3–4cm (1½in) long. In both cases, the layers should be pegged firmly about 7–8cm (3in) deep, after removing any leaves that would otherwise be buried. Pots containing the layers should be kept moist but left undisturbed for at least eighteen months. If roots have not started to grow by the second autumn after layering, they should be induced by partially severing the layered branch just below the surface of the compost.

6
SUMMARY OF PLANTS

	Natural Climate	Form	Page No.
Acacia baileyana	Warm temperate	Shrub	192, *193*
dealbata	Warm temperate	Shrub	191, *192*
longifolia	Warm temperate	Tree	194
melanoxylon	Warm temperate	Tree	194
riceana	Warm temperate	Shrub	194, *195*
Aloe arborescens	Warm temperate	Herb	19, *67*
aristata	Warm temperate	Herb	23
ciliaris	Warm temperate	Herb	21
ferox	Subtropical	Herb	21
mitriformis	Warm temperate	Herb	22
polyphylla	Cool temperate	Herb	20
variegata	Warm temperate	Herb	22
Amaryllis belladonna	Warm temperate	Bulbous herb	23, *25*
Ananas comosus	Tropical	Succulent fruit	135, *137*
Asimina triloba	Cold temperate	Shrub fruit	132
Bougainvillea glabra	Warm temperate	Sprawling shrub	56, *57, 151*
hybrids	Subropical	Sprawling shrub	57
spectabilis	Subtropical	Sprawling shrub	56
Brunfelsia americana	Subtropical	Shrub	141, *142*
calycina	Tropical	Shrub	139, *140*
undulata	Subtropical	Shrub	141
Callistemon citrinus	Warm temperate	Shrub	143, *144*
paludosus	Warm temperate	Shrub	143
phoeniceus	Subtropical	Shrub	143
pithyoides	Cool temperate	Shrub	143
rigidus	Warm temperate	Shrub	143
salignus	Warm temperate	Shrub	143
speciosus	Subtropical	Shrub	143, *144*
subulatus	Cool temperate	Shrub	143
viminalis	Cool temperate	Shrub	143
Carica papaya	Tropical	Tree fruit	130, *131*
Cassia corymbosa	Subtropical	Shrub	147, *148*
didymobotrya	Tropical	Shrub	146, *147*
fistula	Tropical	Tree	147
grandis	Tropical	Tree	147
javanica	Tropical	Tree	147

	Natural Climate	Form	Page No.
Hibiscus rosa-sinensis	Tropical	Shrub	*151*, 179, *181*
sinosyriacus	Warm temperate	Shrub	*181*, 182
syriacus	Cold temperate	Shrub	183
waimeae	Subtropical	Shrub	183
Lagerstroemia indica	Warm temperate	Shrub	*133*, 184
Lapageria rosea	Warm temperate	Twining shrub	*50*, 60
Leucadendron argenteum	Warm temperate	Tree	187
plumosum	Subtropical	Shrub	*188*, 189
stokoei	Subtropical	Shrub	189
Lithops spp.	Subtropical	Succulent herb	*39*, 67
Magnolia grandiflora	Warm temperate	Shrub	73, *75*
Mandevilla suaveolens	Warm temperate	Twining shrub	76, 77
Mimosa pudica	Tropical	Herb	190, *191*
Musa basjoo	Subtropical	Herb	110
cavendishii	Subtropical	Herbaceous fruit	108, *109*
ensete	Subtropical	Herb	110
paradisiaca	Tropical	Herbaceous fruit	108, *109*
Mutisia clematis	Tropical	Climbing shrub	83
decurrens	Warm temperate	Climbing shrub	82
oligodon	Warm temperate	Climbing shrub	*81*, 83
Nerium oleander	Warm temperate	Shrub	197, *199*
Pandorea jasminoides	Subtropical	Twining shrub	84
pandorana	Subtropical	Twining shrub	84
Passiflora antioquiensis	Warm temperate	Climbing shrub	88, 125
caerulea	Warm temperate	Climbing shrub	85, *87*
edulis	Subtropical	Fruiting vine	80, 123, *126*
incarnata	Warm temperate	Fruiting vine	*124*
jamesonii	Subtropical	Climbing shrub	88
laurifolia	Tropical	Fruiting vine	124
ligularis	Tropical	Fruiting vine	125
lutea	Cool temperate	Fruiting vine	124
manicata	Subtropical	Climbing shrub	88
molissima	Subtropical	Fruiting vine	88, 125
quadrangularis	Subtropical	Climbing shrub	87, 125
umbilicata	Warm temperate	Climbing shrub	88
Paullinia thalictrifolia	Tropical	Climbing shrub	89
Petrea kohautiana	Tropical	Twining shrub	*94*
racemosa	Tropical	Sprawling shrub	94
volubilis	Tropical	Twining shrub	*68*, 92
Pharbitis learii	Tropical	Twining shrub	55
Phoradendron flavescens	Cool temperate	Parasitic shrub	78
Phthirusa pyrifolia	Tropical	Parasitic shrub	82
Physalis peruviana edulis	Warm temperate	Herbaceous fruit	121, *122*
Poncirus trifoliata	Cold temperate	Shrub fruit	*116*, 117
Protea barbigera	Cool temperate	Shrub	200
compacta	Warm temperate	Shrub	200

210

	Natural Climate	*Form*	*Page No.*
cynaroides	Warm temperate	Shrub	*67*, 200, *201*
mellifera	Warm temperate	Shrub	200
neriifolia	Warm temperate	Shrub	200
Prunus armeniaca	Cool temperate	Shrub fruit	100, *103*
Psidium guajava	Tropical	Shrub fruit	127, *129*
Rosa banksiae	Warm temperate	Trailing shrub	48, *52*
Schizostylis coccinea	Warm temperate	Herb	29, *30*
Solanum jasminoides	Subtropical	Climbing shrub	90, *91*
melongena esculentum	Warm temperate	Herbaceous fruit	105, *106*
Sollya heterophylla	Subtropical	Twining shrub	53
Strelitzia augusta	Subtropical	Herb	43
reginae	Subtropical	Herb	43, *151*
Streptosolen jamesonii	Subtropical	Sprawling shrub	95, *96, 151*
Telopea mongaensis	Cool temperate	Shrub	204
oreades	Cool temperate	Shrub	204
speciosissima	Warm temperate	Shrub	204, *205*
truncata	Cool temperate	Shrub	*152*, 205
Thunbergia alata	Subtropical	Twining herb	*98*
grandiflora	Subtropical	Climbing shrub	97, *134*
Tropaeolum speciosum	Cool temperate	Climbing herb	70, *71*
Viscum album	Cold temperate	Parasitic shrub	78, *79, 81*
cruciatum	Warm temperate	Parasitic shrub	82
Vitex agnus-castus	Warm temperate	Shrub	155, *156*
Watsonia spp.	Warm temperate	Bulbous herb	45, *46*
Zantedeschia aethiopica	Warm temperate	Tuberous herb	27, *30*

211

CLIMATE MAPS – UK, USA AND CANADA

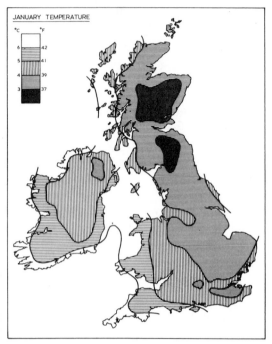

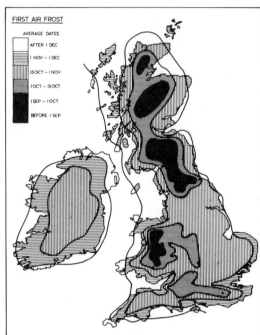

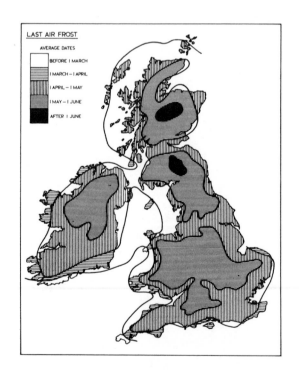

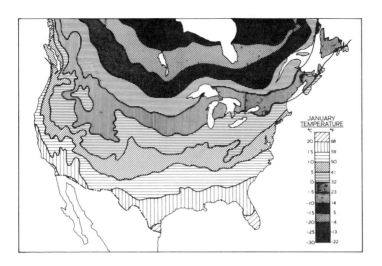

JANUARY
TEMPERATURE

°C		°F
20		68
15		59
10		50
5		41
0		32
-5		23
-10		14
-15		5
-20		-4
-25		-13
-30		-22

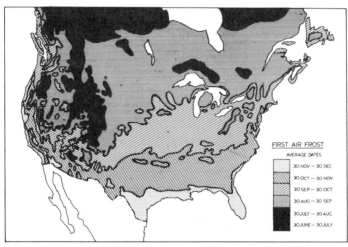

FIRST AIR FROST

AVERAGE DATES

	30 NOV – 30 DEC
	30 OCT – 30 NOV
	30 SEP – 30 OCT
	30 AUG – 30 SEP
	30 JULY – 30 AUG
	30 JUNE – 30 JULY

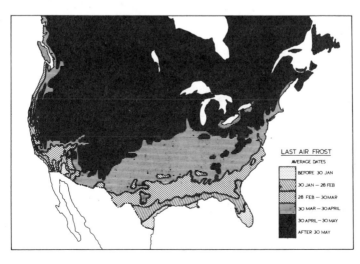

LAST AIR FROST

AVERAGE DATES

	BEFORE 30 JAN
	30 JAN – 28 FEB
	28 FEB – 30 MAR
	30 MAR – 30 APRIL
	30 APRIL – 30 MAY
	AFTER 30 MAY

INDEX

215